Mathematical Foundations and Numerical Analysis of the Dynamics of an Isotropic Universe

The bird-watcher and the cosmic-watcher,
a free interpretation of the contents of this book
by my old school mate Tony Magala (we are the little mice)

Sergio Benenti

Mathematical Foundations and Numerical Analysis of the Dynamics of an Isotropic Universe

Springer

Sergio Benenti
Dipartimento di Matematica
Università di Torino
Turin, Italy

ISBN 978-3-031-64032-2 ISBN 978-3-031-64033-9 (eBook)
https://doi.org/10.1007/978-3-031-64033-9

Translation from the Italian language edition: "Fondamenti matematici e analisi numerica della dinamica di un Universo isotropo" by Sergio Benenti, © Accademia delle Scienze di Torino, 2020. Published by Accademia delle Scienze di Torino. All Rights Reserved.

This Springer imprint is published by the registered company Springer Nature Switzerland AG
The registered company address is: Gewerbestrasse 11, 6330 Cham, Switzerland

Preface

In the preface to his book *A Deductive Theory of Space and Time* (1966) [5], the logician-mathematician S. Basri writes:

> At present Physics is a collection of deductive theories, many of which do not explicitly specify all the concepts and postulates on which they are based.

This criticism is still valid today for cosmological theories, despite the fact that cosmologists found, in the late 1990s, a *cosmological agreement* on a standard cosmological model called ΛCDM-model (Lambda model with cold dark matter). Unfortunately, however, the paper in which this agreement was written and signed seems to be untraceable in the literature. It is like the Arabic Phoenix of Mozartian memory in *Cosí fan tutte*: *che ci sia ciascun lo dice, dove sia nessun lo sa.*

It is on the other hand widely believed that the lack of a logical-deductive structure makes painful and fruitless the reading of most books or articles on cosmological topics, so much so that someone has ironically spoken of '*expanding Universe and expanding confusion*'.

This monograph sets out some of the main results obtained in the course of a research project begun several years ago and aimed at satisfying a personal need: to be able to trace back the most striking results in cosmology, sometimes controversial, to a system of postulates clearly expressed in mathematical terms.

Two methodological principles have been adopted since the beginning of this research: a *principle of simplicity* and a *principle of good ordering*.

Simplicity

An all-encompassing mathematical model capable of describing the complexity of phenomena that occurs during the evolution of the Universe, especially in its primordial stages, is not conceivable by the human mind.

Therefore, we must moderate our ambitions, aiming for simple and meaningful models that will enable us to enter the complex field of cosmology in a comprehensible and fruitful way.

On the other hand, as today we understand well thinking about the historical evolution of geometry, that is, the transition from Euclidean geometry to non-Euclidean geometries, postulates should not be conceived as *absolute acts of faith*. On the contrary, they must be *questionable*. The advantage of an axiomatic formulation of a theory lies precisely in the possibility of identifying the postulates to be modified, or added, in order to get a theory fitting a larger, or different, class of physical phenomena.

Good Ordering

This principle concerns the proper sequencing of postulates and theorems. In mathematics, this principle requires that postulates be placed initially in a general category and later in a chain of subcategories. In cosmology, as well as in classical or relativistic mechanics, the first postulates and theorems must deal with issues of pure geometry and kinematics. These must then be followed by postulates of a dynamic nature.

In compliance with this principle, the postulates are in this memoir divided into three groups.

1. The **first group of postulates** concerns the **geometry of cosmic space-time** (Chap. 1) regardless of the physical phenomenology governing the evolution of the Universe. As will be seen, although based on shareable elementary concepts, these postulates yield a conspicuous mass of results, some known but many others less well known.

Among the most interesting novelties we find that the **scale factor**, also called **expansion factor**, which is the conformal factor between two spatial metrics of the Universe at two different times, turns out to depend on the choice of a **reference time** denoted here as $t_\sharp$. This notion is not only of theoretical interest but also provides a very useful and effective operational tool. Indeed, the scale factor is a dimensionless scalar function of cosmic time t, usually denoted by $a(t)$, from which the evolution of most of the physical quantities in the Universe, and thus the evolution of the entire Universe, can be described qualitatively and in some cases also numerically.

Here it will be denoted by $a(t, t_\sharp)$ to indicate its dependence on the reference time $t_\sharp$. It is clear that, to make geometrical or physical sense, the equations or definitions involving it, possibly together with its derivatives, must be independent of the choice of $t_\sharp$. A simple example is given by the **Hubble factor** that is involved in the famous **Hubble law** and that in our context is defined by

$$H(t) = \frac{\dot{a}(t, t_\sharp)}{a(t, t_\sharp)}.$$

Despite the fact that the reference time $t_\sharp$ appears on the right-hand side, calculations show that the fraction is actually independent of it, so $t_\sharp$ need not appear on the left-hand side.

Perhaps this is why cosmologists did not feel the need to introduce the notion of **reference time**, partly because in the current common notation $a(t)$ it is tacitly assumed that $t_\sharp$ is the present time t_0, so that $a(t_0) = 1$.

A second novelty is the theorem of existence in cosmic space-time of symmetric linear connections that we call **cosmic connections**, satisfying certain compatibility requirements with the geometrical structures already present in space-time as a consequence of the first set of postulates.

Connections are a necessary prelude to the formulation of dynamical theories because they allow for the extension of the notions of **acceleration** and **derivative of vector fields** from affine spaces to differentiable manifolds.

It is shown, *a remarkable fact*, that every cosmic connection turns out to be determined by a single function of cosmic time t, denoted here by $F(t)$.

2. The **second group** of **bridge-postulates** (Chap. 2) concerns the transition from geometry to dynamics. With the choice of a bridge-postulate, one moves into the territory of dynamics where, in order to continue on the path, it is necessary to introduce dynamic postulates (the **third group of postulates**). A bridge-postulate is acceptable and effective if it allows to determine a single function $F(t)$, thus a single cosmic connection (Fig. 1).

Two bridge-postulates are proposed in Chap. 2. The first one leads to Newtonian cosmological models, but, after a brief discussion, we do not continue further in this direction. Instead, the second one leads to the dominion of **relativistic cosmological models**. It assumes the existence of a **cosmic time** t and of special **wandering particles** whose **peculiar velocity** (Sect. 2.2) is always and everywhere equal to a universal constant c. These particles are given the name **photons**. From this postulate, it follows not only the existence of a special cosmic connection but also that this derives from a Lorentzian metric (i.e. it is a Levi-Civita connection) for which the world-lines of photons are null (light-like) while the world-lines of the galactic fluid are time-like. As a consequence we find, as a theorem, the **Weyl principle** (Sect. 1.3).

3. With this last bridge-postulate, one moves to the **relativistic cosmic dynamics** (Chap. 3) based on three dynamical postulates.

3.1. With the **first relativistic dynamical postulate**, we assume the Einstein equations as field equations. Already in Chap. 2, the ingredients needed to compose these equations were prepared, namely the Ricci $R^{\alpha\beta}$, and Einstein $G^{\alpha\beta}$, tensors associated with the relativistic cosmic connection (Sect. 2.5).

But even earlier, in Chap. 1, it was seen that, by virtue of the principle of isotropy, every symmetric two-tensor $T^{\alpha\beta}$ turns out to be completely determined by only two characteristic functions $\phi(t)$ and $\psi(t)$ of the cosmic time, and it is shown that the four conservation equations $\nabla_\alpha T^{\alpha\beta} = 0$ reduce to a single ordinary first-order differential equation in the functions ϕ and ψ, *whatever the cosmic connection.*

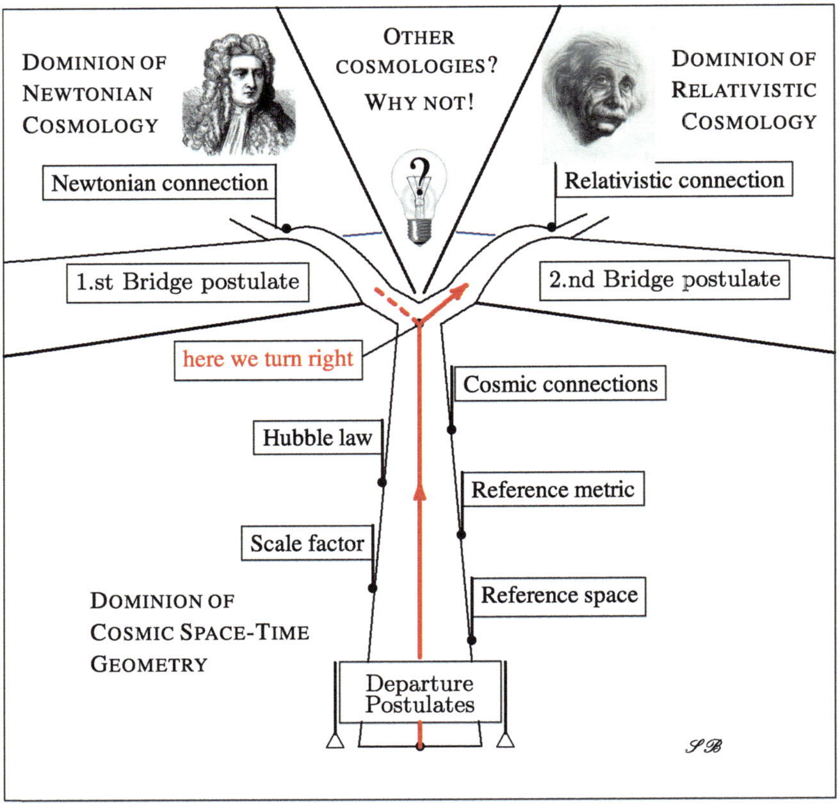

Fig. 1 Our cosmic route

By virtue of this general property, it is shown (Theorem 3.1) that Einstein's ten equations are in fact equivalent to only two differential equations (of first and second order) in the scale factor $a(t, t_\sharp)$ involving the two characteristic functions ϕ and ψ of the energy tensor, the cosmological constant Λ, and the spatial curvature constant $\widetilde{K}$ at the reference time $t_\sharp$.

These dynamical equations are similar to the celebrated **Friedmann-Lemaître-Robertson equations**, but, unlike these, our equations are valid in all generality for any kind of energy tensor. In this regard, it was deemed appropriate to carry out a detailed examination of the Friedmann and Lemaître equations (Sect. 3.4) in order to solve some interpretation issues.

3.2. With the **second relativistic dynamical postulate** (Sect. 3.2), the energy tensor of the cosmic fluid is assumed to be that of a perfect fluid:

$$\boxed{T^{\alpha\beta} = c^{-2}(\varepsilon + p)V^\alpha V^\beta + pg^{\alpha\beta}}$$

where V^α is the **absolute velocity** of the cosmic fluid, for which the normalization condition $g_{\alpha\beta}V^\alpha V^\beta = -c^2$ applies, $\varepsilon(t)$ is the **energy density**, and $p(t)$ is the **pressure**. It is then shown that the characteristic functions are given by

$$\phi = \varepsilon(t), \quad \psi = a^{-2}(t)p(t).$$

3.3. Finally, a **third relativistic dynamical postulate** (Sect. 3.3) concerns the **equations of state,** i.e., the kinds of functions $\varepsilon(t)$ and $p(t)$ and the relationships to be prescribed for them. The total energy density $\varepsilon(t)$ is viewed as the sum of the energy densities $\varepsilon_i(t)$ of various components of the cosmic fluid related to the corresponding pressures p_i by the simple linear relations $p_i = w_i\varepsilon_i$ where w_i are dimensionless constants called the bf parameters of state. Hence, on this postulate, possibly implemented, one can base the so-called **many-component cosmological models**. However, these models are unreliable, partly because of their complexity.

4. At this point the principle of simplicity induces us to take into account the two fundamental types of energy distributed in the Universe, *matter energy* and *radiation energy*, although it is customary to add a third one, that of *dark energy*, to which, however, we shall give here a distinguished role.

This choice is formalized in a **fourth postulate** (Sect. 3.5): the energy density ε present in the Universe is the sum

$$\boxed{\varepsilon = \varepsilon_m + \varepsilon_r}$$

of a **matter energy density** and a **radiation energy density** with two distinct characteristic properties. The density ε_m does not generate pressure

$$\boxed{p_m = 0 \ \ (w_m = 0)}$$

and is itself the sum

$$\boxed{\varepsilon_m = \varepsilon_b + \varepsilon_c}$$

of an **energy density of baryonic matter** and of an **energy density of cold dark matter** whose nature is so far rather unknown. Instead, the radiation density ε_r generates a pressure with equation of state

$$\boxed{p_r(t) = \frac{1}{3}\varepsilon_r(t) \ (w_r = \frac{1}{3})}$$

As a **fifth postulate,** we assume that there exists an **equaliztion time** t_{eq} (also called **balancing time**) at which the energy densities of matter and radiation take the same value:

$$\boxed{\varepsilon_m(t_{\text{eq}}) = \varepsilon_r(t_{\text{eq}})}$$

It should be emphasized that all these assumptions are supported by commonly accepted arguments reported in astrophysics texts.

5. On this basis we construct a cosmological model which we call the **MR-model** (matter-radiation-model) to which almost all the rest of the work is devoted. This model differs from the standard ΛCDM-model in that the dynamical equation governing the evolution of the scale factor $a(t, t_0)$, with reference time the present time t_0, can be reduced to the form (Theorem 3.6)

$$[*] \quad \boxed{\dot{a}^2 = H_0^2\left[1 + \Omega_\Lambda\left(a^2 - 1\right) + \Omega_m\left(a^{-1} - 1 + \frac{a^{-2} - 1}{1 + z_{\text{eq}}}\right)\right]}$$

where only four constants are involved:

$$\begin{cases} H_0 & \text{current value of the Hubble factor,} \\ \Omega_\wedge & \text{dark energy parameter,} \\ \Omega_m & \text{matter energy parameter,} \\ z_{\text{eq}} & \text{redshift corresponding to time } t_{\text{eq}}. \end{cases}$$

We will call them **primary data**. Of the four **cosmological parameters** of the ΛCDM-model,

$$\left[\begin{array}{l} \text{dark energy, } \Omega_\Lambda \overset{\text{def}}{=} \frac{1}{3}\frac{c^2}{H_0^2}, \Lambda \Longleftrightarrow \Lambda = \frac{3H_0^2}{c^2}\Omega_\Lambda, \\ \text{matter, } \Omega_m \overset{\text{def}}{=} \frac{1}{3}, \frac{c^2}{H_Z^2}, \chi\ \varepsilon_{m0} \Longleftrightarrow \chi\ \varepsilon_{m0} = \frac{3H_0^2}{c^2}\Omega_m, \\ \text{radiation, } \Omega_r \overset{\text{def}}{=} \frac{1}{3}, \frac{c^2}{H_Z^2}, \chi\ \varepsilon_{r0} \Longleftrightarrow \chi\ \varepsilon_{r0} = \frac{3H_0^2}{c^2}\Omega_r, \\ \text{curvature, } \Omega_K \overset{\text{def}}{=} \Omega_\Lambda - \Omega_m - 1, \end{array}\right.$$

only Ω_Λ and Ω_m remain, while the other two, Ω_r and Ω_m, are replaced by the redshift z_{eq} corresponding to the balancing time t_{eq} between the matter and radiation densities. This reduction is possible thanks to the fact that the existence of the equalization time t_{eq} implies a linear relationship between the matter parameter and the radiation parameter:

$$\Omega_r = \frac{\Omega_m}{1 + z_{\text{eq}}}.$$

A second non-negligible advantage of the MR-model is that the spatial curvature parameter Ω_K does not enter the dynamical equation [*]. This avoids falling into the error of assuming that the curvature is null (*flatness assumption*) simply because from observational data its value turns out to be extremely small. In fact, it is possible to prove (Theorem 3.7) that *regardless of the values of the primary data, in the MR-model the spatial curvature cannot be zero.*

6. The analysis of the dynamical equation [$*$] is made easy by the fact that it is a **Weierstrass equation**, i.e., of the type

$$\dot{a}^2 = W(a).$$

Indeed, such an equation offers three significant advantages.

(i) Even if one does not know how to integrate it, looking at the graph of the **Weierstrass function** $W(a)$ in the plane ($x = a, y = \dot{a}^2$) one can recognize the main properties of the solutions $a(t)$.

(ii) The integral

$$[**]\ t(a) = \int_0^a \frac{dx}{\sqrt{W(x)}}$$

gives the time t at which the scale factor takes the value a, the upper bound of the integration interval. Therefore, calculating these integrals for a dense sequence of values of a, starting with $a = 0$, and reversing the tabulation of the function $t(a)$ so obtained we get a **pointwise profile** of the Universe $a(t, t_0)$ that starts from the origin $(0, 0)$ and thus predicts the existence of the singularity $a = 0$ for $t = 0$, the so-called **big-bang**. In particular, for $a = 1$ the integral [$**$] gives the **age of the Universe**:

$$\boxed{t_0 = \int_0^1 \frac{dx}{\sqrt{W(x)}}}$$

The graphs in the (t, a) plane of the solutions will be called **profiles of the Universe**.

(iii) With respect to the ordinary *step-by-step* numerical integration methods, we can better take under control the error in calculating $t(a)$.

7. With Chap. 4 the numerical analysis begins. It is first necessary to fix the primary data values to be fed in the dynamical equation [$*$]. The analysis of the numerous

data reports associated with the various space projects carried out in recent years, briefly mentioned in Sect. 4.1, leads to consider sufficiently reliable the estimates of the following **primary data** (Table 1):

Table 1 Primary data for the MR-model

$\widehat{H}_0$	70.0 $\mathrm{kms}^{-1}\ \mathrm{Mpc}^{-1}$	Ligo [14]
$\widehat{H}_0$	67.74 $\mathrm{kms}^{-1}\ \mathrm{Mpc}^{-1}$	Planck [2]
Ω_Λ	0.6911	Planck [2]
Ω_m	0.3089	Planck [2]
z_{eq}	3371	Planck [2]

With these estimates it can be shown that *according to the MR-model, the spatial curvature is positive* (Theorem 4.1) and one can calculate its present values (Sect. 4.5)

$$\widehat{H}_0 \mapsto \boxed{K_0 \simeq 0.46885 * 10^{-6}\, Glyr^{-2}} \quad \bar{H}_0 \mapsto \boxed{K_0 \simeq 0.43907 * 10^{-6}\, Glyr^{-2}}$$

together with the curvature radii

$$\hat{H}_0 \mapsto \boxed{r_0 \simeq 1460.4299\, Glyr} \quad \bar{H}_0 \mapsto \boxed{r_0 = 1509.1540\, Glyr}$$

Regarding the Hubble constant H_0, we take into account two estimates. The first, denoted by $\widehat{H}_0$, is recently acquired and considered to be of high precision. It was provided by the 2017 LIGO gravitational wave experiment. The second, denoted by $\widehat{H}_0$, dates back to the 2016 Planck project report.

The reason for this double choice is because the (square of the) Hubble constant H_0 is a factor of the Weierstrass function in the dynamical equation [$*$] so that the dynamics is significantly affected by even small changes in H_0.

This 'sensitivity' then propagates to all numerical data derived from $W(a)$. For example, for the age of the Universe we obtain the two estimates

$$\hat{H}_0 \mapsto \boxed{t_0 \simeq 13.36116\, Glyr} \quad \bar{H}_0 \mapsto \boxed{t_0 = 13.80692\, Glyr}$$

Time t_0 is one of the **four key-times** t_* that we have to consider because of the crucial role they will play. To these key-times correspond as many key-values a_* of the scale factor as shown in the following Table 2:

Table 2 Key-times t_* and key-values a_*

a_*	Event	t_*
1	Present time	t_0
a_{eq}	Balance of matter and radiation density	t_{eq}
a_q	Beginning of accelerated expansion	t_q
a_{re}	Reionization (beginning of light emission)	t_{re}

The decrease of H_0 in the transition from $\widehat{H}_0$ to $\widehat{H}_0$ has the effect of increasing all dates and thus moving the profile of the Universe toward the future, as shown in Fig. 2. Given the values of the key-scale factors a_*, the corresponding dates are calculated with the usual integral

$$t_* = \int_0^{a_*} \frac{dx}{\sqrt{W(x)}}.$$

The resulting estimates are (Sect. 4.8)

$\widehat{H}_0 \mapsto t_{\mathrm{eq}} \simeq 50,1595\,yr$	$\widehat{H}_0 \mapsto t_{\mathrm{re}} \simeq 0.54409\,Gyr$
$\bar{H}_0 \mapsto t_{\mathrm{eq}} \simeq 51,8330\,yr$	$\bar{H}_0 \mapsto t_{\mathrm{eq}} \simeq 0.56224\,Gyr$

$\widehat{H}_0 \mapsto t_q \simeq 7.37949\,Gyr$
$\bar{H}_0 \mapsto t_q \simeq 7.62569\,Gyr$

The determination of a numerical profile of the MR-model is in itself a significant achievement, but what we actually need to continue our analysis is an analytical representation of it. However, it is unthinkable to search for an exact solution of the Weierstrass equation. It would also be futile because, in all probability, it would involve non-elementary transcendental functions tractable only through approximate representations.

Fortunately, this crucial problem is solved in Sect. 4.9 by noting that a profile of the type

$$a(t, t_0) = \alpha\sqrt[3]{\cosh(\beta t) - 1}$$

with α and β positive constants is very similar to that obtained numerically. Indeed, it becomes virtually indistinguishable (especially at the key-points mentioned above) if the constants take the values (Theorem 4.3)

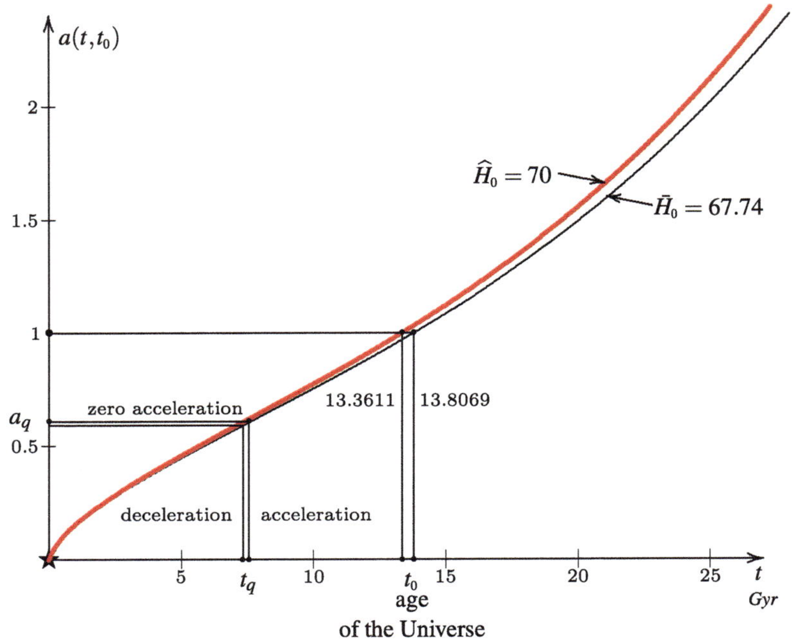

Fig. 2 Profiles corresponding to the two values $\widehat{H}_0$ and $\bar{H}_0$ of the Hubble constant

$$\alpha \simeq 0.607247, \quad \beta = 0.178366Gyr^{-1},$$

that follow from taking the estimate $\widehat{H}_0$ for Hubble's constant.

These profiles are in perfect agreement with Fig. 3 taken from A. G. Riess' Nobel Lecture and conveniently reworked.[1] The envelope of the spatial sections (red curve) has the same trend as the scale factor.

Given this analytical profile, a theoretical and numerical analysis concerning the transmission of light signals, the visibility of the Universe, and the redshift phenomenon is performed in the last two chapters. The main results obtained can be summarized in Fig. 4, which provides an overview of the configuration of the Universe referring to the present epoch and to an observer located in any B galaxy.

The great circle represents the three-dimensional sphere $\mathbb{S}_3$ where the galaxies are currently distributed. Its radius is $r_0 \simeq 1460, 42\,Glyr$ (4.15). The upper part of the figure shows how the observer in B sees the cosmos in its vicinity. Because of the extremely small curvature of $\mathbb{S}_3$, the Universe appears to B flat at least up to a distance of about 29.59 billion light-years. This is the **current visibility radius** $r_{vis}(t_0)$ beyond which nothing can be seen today. There is also an **absolute invisibility radius of the Universe**

[1] 2011 Nobel Prize in Physics, together with Saul Perlmutter and Brian P. Schmidt '*for the discovery of the accelerating expansion of the Universe through observations of distant supernovae*'.

Fig. 3 Riess profile

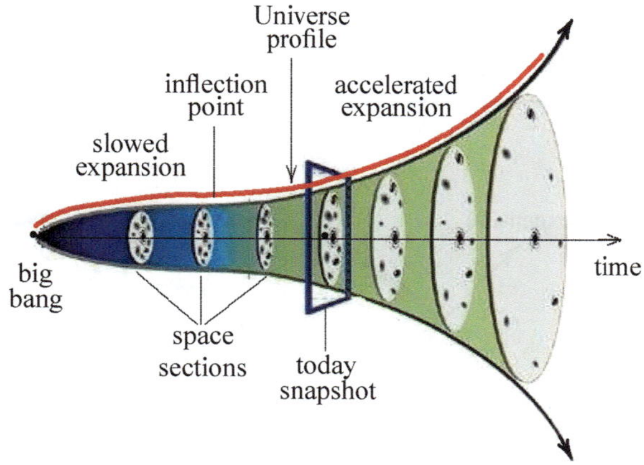

$$r_{\text{inv}} \simeq 45.61\,Glyr$$

beyond which the Universe remains forever invisible to B.

These two '*radii*' are in fact distances measured on the sphere $\mathbb{S}_3$, starting from B, thus subtending two angles at the center $\psi_{\text{vis}}(t_0)$ and ψ_{inv}. In the course of time, the current visibility angle $\psi_{\text{vis}}(t_0)$ tends asymptotically to the absolute invisibility angle ψ_{inv}, which instead remains constant. It should be mentioned that they are not shown in the same scale of the rest of the figure. In fact, respecting the scale they would turn out to be almost imperceptible because their value in radians is very small:

$$\psi_{\text{vis}}(t_0) = \frac{r_{\text{vis}}(t_0)}{r_0} \simeq 0.020262, \; \psi_{\text{inv}} = \frac{r_{\text{inv}}}{r_0} \simeq 0.031233.$$

The conclusion is astonishing:

Suppose we live in the galaxy B. There is a world of galaxies immensely larger than that we can observe today and that will forever remain unknown to us. However, this unseen world affects the evolution of the entire Universe, thus also that part visible to us today.

But there is still something interesting we must realize:

Think of an observer placed in any galaxy B′ of that world today invisible to us. He will arrive at the same geometrical and numerical conclusions represented in Fig. 4.

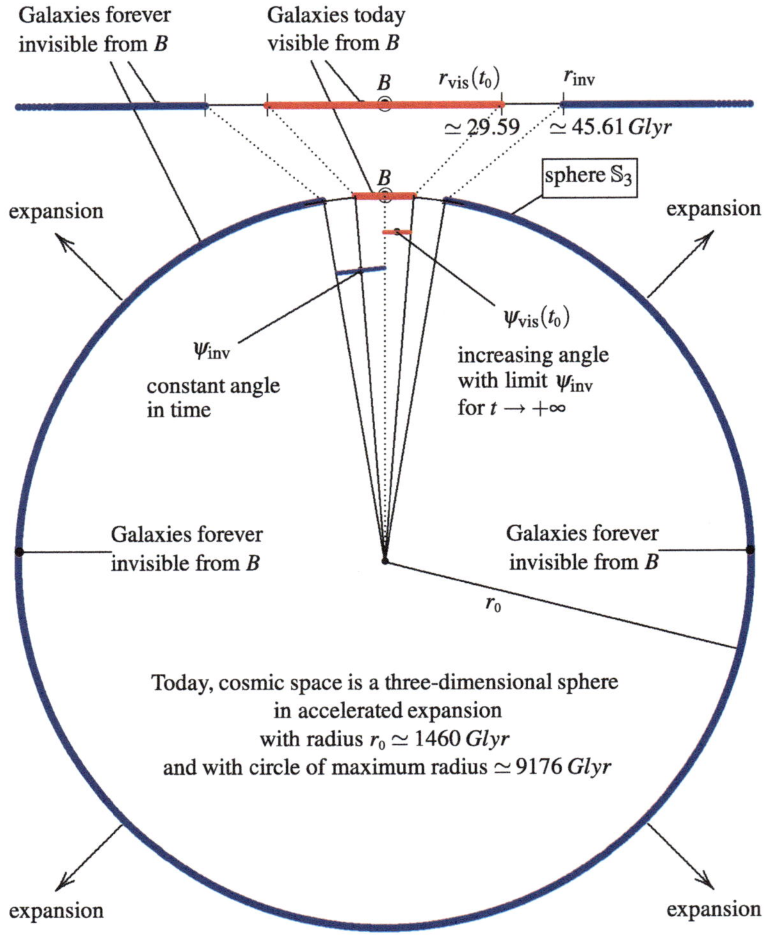

Fig. 4 Present time Universe configuration referred to an observer placed in a galaxy B

Known and lesser-known results can be found in this book,
but in any case the method by which they are obtained
is definitely new.

Proposal

Even if we believe that the Universe is printed in a single copy we cannot hope to set up a unique mathematical model capable of describing its evolution in time with the desired accuracy. The complexity of the phenomena in the early universe and the high number of observational parameters that can be involved make it impossible to achieve that purpose.

Over the last century, several models have been proposed, modified, and then abandoned. The debate on the involvement of the cosmological constant is a striking example of how opinions change following the astronomical discoveries. Another example is the question about the spatial curvature, erroneously considered to be zero by many cosmologists only because the processing of the observational data leads to the conclusion that it is negligible. On the other hand, too many are the models that nowadays are still proposed and analyzed, so a certain confusion has been created, especially because for most of them it is not clear on which general principles (or postulates) they are founded.

The standard approaches to cosmology are based on Weyl's principle and the cosmological principles of homogeneity and isotropy. But by following this shortcut, we lose the important occasion of being able to distinguish if a certain property has a purely geometrical character or depends on the field equations.

The purpose of this monograph is to demonstrate how to bring order to this complex subject, starting from 'principles' or 'postulates' declared in understandable mathematical terms. Once subscribed, these postulates open the way to a series of theorems. If you do not agree with some of them, then you have to re-tune the postulates or, in the extreme case, look for another approach.

Declaration This monograph is an improved and expanded English version of a Memoir entitled *Fondamenti matematici e analisi numerica della dinamica di un Universo isotropo* published by Accademia delle Scienze di Torino in the volume no. 42–43, years 2018–2019.

Contents

1 Geometry of the Cosmic Space-Time 1
1.1 Geometrical Postulates 1
1.2 The Isotropy Principle and Its Consequences 3
1.3 Comments on the Weyl Principle 5
1.4 Linear Connections and Curvature 5
1.5 Permanence of the Sign of Spatial Curvature 9
1.6 Conformal Factor Between Two Spatial Metrics 11
1.7 Reference Time and Scale Factor 12
1.8 Recession Speed and Hubble Law 14
1.9 Dimensional Analysis 15
1.10 Co-moving Coordinates 17
1.11 Isotropic Vectors and Tensors 18
1.12 Cosmic Monitor and Wandering Particles 20
1.13 Cosmic Connections 22
1.14 Ricci Tensor of a Cosmic Connection 30
1.15 Covariant Derivatives and Conservation Equations 30

2 Bridge-Postulates 33
2.1 Newtonian Cosmic Connection 33
2.2 Peculiar Velocity of a Wandering Particle 35
2.3 Relativistic Cosmic Connection 36
2.4 Relativistic Cosmic Metric 40
2.5 Ricci and Einstein Tensors 43
2.6 Sub-luminal Particles 45

3 Relativistic Cosmic Dynamics 47
3.1 First Dynamical Postulate: Einstein Equations 47
3.2 Second Dynamical Postulate: Perfect Cosmic Fluid 49
3.3 Third Dynamical Postulate: State Equations 51
3.4 Comments on Friedmann and Lemaître Equations 52
3.5 Fourth Dynamical Postulate: Matter and Radiation 55
3.6 Dynamical Equations of the MR-Model 57
3.7 Relationship Between Matter and Radiation Densities 58
3.8 The Problem of the Spatial Curvature 60
3.9 Weierstrass Equation 62
3.10 Profiles of the Universe 64
3.11 Qualitative Profiles of the MR-Model 65
3.12 Complements on the Weierstrass Equation 67

4 Numerical Cosmology 77
4.1 Gathering Cosmological Data 77
4.2 In the MR-Model the Spatial Curvature Is Positive 79
4.3 Closed Universe Model 80
4.4 Measurement Units of the Hubble Factor 82
4.5 Space Curvature Estimates 83
4.6 Weierstrass Functions of the MR-Model 84
4.7 Pointwise Numerical Profiles of the MR-Model 84
4.8 Age of the Universe and Other Key-Dates 87
4.9 Analytical Profile of the MR-Model 89
4.10 Changing the Reference Time 92
4.11 Time Evolution of the Hubble Factor 93
4.12 Super-Luminal Recession Speed 94
4.13 Models with Null Cosmological Constant 97

5 Signal Transmission and Visibility 101
5.1 Photon Transmission 101
5.2 Emission-Reception Relationship 102
5.3 Visible and Unvisible Universe 106
5.4 Present-Time Configuration of the Universe 110
5.5 Visibility and Super-Luminal Recession Speed 112

6 Cosmic Redshift 115
6.1 Cosmic Redshift 115
6.2 From Redshift to the Current Distance 122
6.3 From Redshift to the Distance at the Emission Time 122
6.4 Slipher Law and Constant Expansion Models 124

7 Appendices 131
7.1 Stereographic Projection of Hyper-spheres 131
7.2 Volume of a Hyper-spherical Cap 134

Corretion to: Numerical Cosmology C1

References 139

Index 141

Chapter 1
Geometry of the Cosmic Space-Time

1.1 Geometrical Postulates

In a simplified, large-scale view of cosmology, we can think of the bodies spread throughout the Universe as particles constituting a **cosmic fluid**. It is not necessary to discuss now the nature of these **cosmic bodies**; they can be stars, galaxies or whatever. For simplicity's sake, we will interpret them as **galaxies**,[1] The cosmic fluid will therefore also be referred to as **galactic fluid**.

> **First postulate**. *The evolution of the Universe is described in a four-dimensional M manifold called* **cosmic space-time**. *It is made of* **points** *called* **events**.

> **Second postulate**. *The life of a cosmic body (galaxy) is a sequence of events constituting a regular curve in M that we call* **history** *or* **world-line**. *The cosmic fluid histories form a* **congruence of curves** *that fills the entire space-time M* (see Fig. 1.1).

A *congruence* is a bundle of regular curves that never intersect. *Collision between cosmic bodies is therefore excluded by this postulate.*

> **Third postulate**. *In M there exists a foliation of tridimensional submanifolds S_t, called* **spatial sections**, *transversal to the cosmic fluid histories and parameterized by a real number t varying in an open interval $(t_\alpha, t_\omega) \subseteq \mathbb{R}$* (see Fig. 1.2).

The term **foliation** stands for a partition of a manifold M into submanifolds of lower dimension and disjoint, i.e., without points of intersection. A congruence of curves, see above, is an example of foliation where the submanifolds are curves

[1] Let us keep in mind that one of the first fundamental phenomena observed in cosmology is the **Hubble law** which concerns the distance between galaxies and their recession speed.

S. Benenti, *Mathematical Foundations and Numerical Analysis of the Dynamics of an Isotropic Universe*, https://doi.org/10.1007/978-3-031-64033-9_1

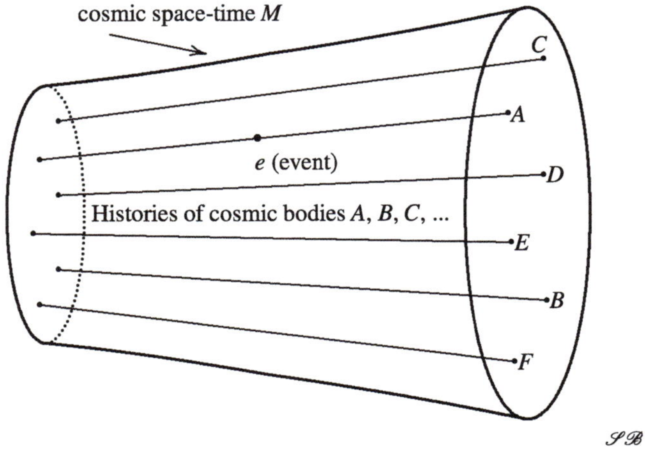

Fig. 1.1 Cosmic space-time M and congruence of cosmic body histories

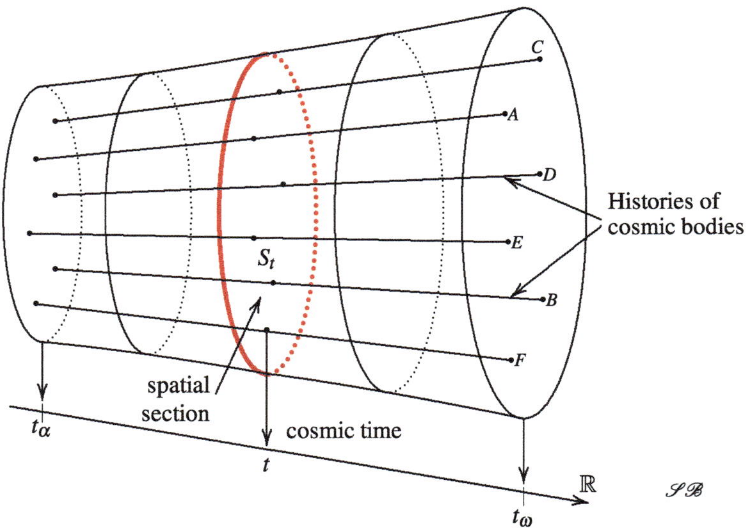

Fig. 1.2 Cosmic fluid hystories, spatial sections and cosmic time

(dimension 1). Furthermore, the term **transversal** means that the histories are never tangent to the spatial sections. The attribute *'spatial'* assigned to sections S_t will be justified later with the fifth postulate.

A first theorem follows, whose proof requires the knowledge elementary notions of differential geometry.

Theorem 1.1 (i) *Cosmic fluid histories establish diffeomorphisms between spatial sections.* (ii) *These stories are themselves diffeomorphic to the open interval* (t_α, t_ω).

The parameter t can be interpreted as **cosmic time**. If an event $e \in M$ is localized in a spatial section S_t then we say that it **happens at time** t, or that **its date is** $t(e)$.

The cosmic time t determines a **chronology of events**: given two events e_1 and e_2, we say that

$$\begin{aligned} e_1 \text{ occurs before } e_2 &\iff t(e_1) < t(e_2), \\ e_1 \text{ occurs after } e_2 &\iff t(e_1) > t(e_2), \\ e_1 \text{ and } e_2 \text{ are simultaneous} &\iff t(e_1) = t(e_2). \end{aligned}$$

Therefore the spatial sections S_t are composed of **simultaneous events**.

For now, cosmic time t is not uniquely determined. It can be replaced by any other parameter, as long as the latter does not change the chronology, which thus has an absolute character. Time t will acquire physical meaning only through a bridge-postulate (Chap. 2) preparatory to the dynamics.

The establishment of an absolute time, which is one of the postulates of Newtonian mechanics, may raise some perplexity. However, it is a necessary prelude to the definition of the concepts of **homogeneity** and **spatial isotropy**, which we shall see in a moment and which together form the so-called **cosmological principle** adopted in most cosmological models.

On the other hand, the existence of submanifolds of simultaneous events is justified by the assumption (which will result in a postulate) that there exist scalar physical entities $\rho(t)$, such as the density of matter or radiation, which are monotonic functions of time, i.e., always increasing or always decreasing. It follows that two events have to be considered simultaneous if the density rho takes the same value in them, and thus that the density ρ can be taken as the time parameter at the place of t.

1.2 The Isotropy Principle and Its Consequences

The **Copernican principle** assumes that neither the Sun nor the Earth are in a special, particularly favored position in the Universe. We can strengthen this principle by requiring that there are no favored directions either. This is the so-called **principle of isotropy**. Since we can interpret a spatial cross section S_t as a *snap-shot* of the three-dimensional physical world at time t, we can express the principle of isotropy as follows:

> **Forth postulate.** *On each space section S_t there are no privileged vector fields having geometrical or physical meaning.*

The tangent vectors to the space sections will be said to be **space vectors**. It is reasonable to think that in each of the snap-shots of the physical world we can measure lengths, angles, volumes, etc. We translate this thought into the following postulate.

Fifth postulate. *In cosmic space-time there exists a tensor field g, covariant and symmetrical, which cancels on the vectors tangent to the curves of the cosmic fluid and whose restriction on each space section S_t defines a positive metric tensor g_t.*

Remark 1.1 In this postulate it is understood that g, and hence the tensors g_t, are of sufficiently high class to guarantee the validity of the formulas in which they will be involved. In any case, there is no loss of generality in considering of class C^∞ all the scalar, vector and tensorial fields introduced so far and which we will introduce in the following. •

Each space section S_t thus turns out to be a three-dimensional Riemannian manifold with positive definite metric (see Fig. 1.2).

By means of the metric g_t we can define the **gradient** of any scalar field over S_t. Such a gradient is a privileged vector field. This is contrary to the isotropy principle. The following theorem avoids its existence.

Theorem 1.2 **Spatial homogeneity**: *from the isotropy principle it follows that every scalar field on M having geometric or physical meaning is a function of cosmic time t only, i.e., it is constant on every spatial section S_t.*

Spatial homogeneity is also named **spatial uniformity**.

In the axiomatic path we are following, spatial homogeneity is thus a consequence of isotropy, whereas in the classical approach to cosmology homogeneity is considered as a 'principle' along with isotropy (see above). In fact, as shown by the following theorem, the isotropy principle implies the Copernican principle mentioned at the beginning of this section.

Theorem 1.3 *The isotropy principle implies the non-existence on any spatial section S_t of privileged points having a geometrical or physical significance.*

Trace of the proof. In a (sufficiently restricted) neighborhood $U \subset S_t$ of such a point P we can define, by means of the metric g_t, the distance $d_{P,X}(t)$ of each point $X \in U$ from P. Consequently, the gradient of $d_{P,X}$ results in a privileged vector field on U, in contrast with the isotropy principle. ∎

Theorem 1.4 *Each space section (S_t, g_t) is a three-dimensional Riemannian manifold with constant curvature.*

Trace of the proof. Here basic notions of Riemannian geometry are required, partly recalled in the following Sect. 1.4. In any three-dimensional Riemannian manifold the Ricci tensor determines three privileged directions at each point, against to the isotropy principle. These directions are not defined if and only if the Ricci tensor is proportional to the metric tensor: $R_t = \lambda_t\, g_t$. The factor λ_t must be constant on S_t by virtue of the Theorem 1.2. Thus, (S_t, g_t) turns out to be an **Einstein manifold**. A theorem of Riemannian geometry states that every three-dimensional Einstein manifold has constant curvature. ∎

1.3 Comments on the Weyl Principle

Generally, cosmology treatises consider the Weyl principle and the cosmological principle (isotropy and homogeneity) as bases for constructing models of the dynamics of the Universe. They are commonly formulated as follows (see for instance [17]).

Weyl principle: *In cosmic spacetime the world-lines of the galaxies form a bundle of non-intersecting time-like geodesics orthogonal to a series of space-like hypersurfaces.*

Cosmological principle: *On large scales the Universe is spacely homogeneous and spacely isotropic.*

First Comment. Weyl's principle places cosmology in the relativistic domain from the very beginning. In our approach, the second part of this principle, concerning geodesics, is a theorem (Theorem 2.9) placed in Chap. 2 (bridge-postulates) while the first part concerning cosmic histories is part of our second postulate. One may ask the question: why not accept Weyl's principle from the beginning instead of spending so much time and space starting from several much weaker postulates? The answer is that, first, our postulates are easy and straightforward to understand for those who are not readily familiar with the theory of general relativity. The second reason is that by following our longer way we do not lose sight of interesting and important facts that are not strictly relevant to relativity. It is the difference between a trip by car and a trip by plane. With a car one can closely observe the succession of picturesque sights and splendid monuments. Apart from this it should be noted that, as we will see, Hubble's law as well as many other concepts, concerning for example the scale factor and reference space, isotropic tensors and cosmic connections, which are independent of any assumptions about the 'physics' of the cosmic fluid, would be left forever in darkness.

Second comment. About the cosmological principle it should be noted that in our approach the principle of isotropy implies (as a theorem) spatial homogeneity.

1.4 Linear Connections and Curvature

Let us recall some basic notions of differential geometry, not only to clarify the meaning of what was said in the preceding sections but also because we will refer to them often in what follows. Consider an n-dimensional variety Q with local coordinates $(q^\alpha) = (q^1, \ldots, q^n)$.

1. A **connection Γ on a manifold Q** is a **rule**, called **parallel transport**, according to which, assigned on Q any parameterized curve $q^\alpha(\xi)$ on an interval (ξ_0, ξ_1) and given any vector v_0 at the initial point $P_{\xi_0} = [q^\alpha(\xi_0)]$, a vector v_ξ at every point $P_\xi = [q^\alpha(\xi)]$ of the curve and thus also at the final point $P_{\xi_1} = [q^\alpha(\xi_1)]$, results to be uniquely determined. It should be said immediately that, keeping the initial and final points as well as the vector v_0 fixed, but changing the curve joining these points,

the vector carried to the final point is, in general, different from the vector obtained by traversing the previous curve.

2. A connection is said to be **linear** if its parallel transport commutes with the linear combination of vectors. In the domain of a given coordinate system $(q^\alpha) = (q^1, \ldots, q^n)$ a linear connection Γ is represented by three-indexed symbols $\Gamma^\gamma_{\alpha\beta}$, functions of the coordinates, such that the transport of a vector $v^\alpha(\xi)$ along a curve $q^\alpha(\xi)$ is governed by the **transport equations**

$$\frac{dv^\gamma}{d\xi} + \Gamma^\gamma_{\alpha\beta}\, v^\alpha \frac{dq^\beta}{d\xi} = 0, \tag{1.1}$$

From now on we will consider only linear connections. The 'linear' attribute will almost always be omitted. By changing the coordinate system, the symbols of a connection change according to the law

$$\Gamma^\gamma_{\alpha\beta} = J^\gamma_{\gamma'}\, \Gamma^{\gamma'}_{\alpha'\beta'}\, J^{\alpha'}_\alpha\, J^{\beta'}_\beta + J^\gamma_{\gamma'} \partial_\alpha J^{\gamma'}_\beta, \quad J^{\alpha'}_\alpha \stackrel{\text{def}}{=} \frac{\partial q^{\alpha'}}{\partial q^\alpha}, \quad J^\alpha_{\alpha'} \stackrel{\text{def}}{=} \frac{\partial q^\alpha}{\partial q^{\alpha'}}. \tag{1.2}$$

which gives **intrinsic meaning** to the transport Eq. (1.1).[2]

3. A connection is said to be **symmetric** if its symbols are symmetrical in the lower indices, $\Gamma^\gamma_{\alpha\beta} = \Gamma^\gamma_{\beta\alpha}$. It can be shown that this property does not depend on the choice of coordinates.

4. A symmetric connection determines two **curvature tensors** of fundamental importance, the **Riemann tensor** and the **Ricci tensor**. Their components are respectively defined as follows[3]:

$$\boxed{R^\nu{}_{\alpha\mu\beta} \stackrel{\text{def}}{=} \partial_\mu \Gamma^\nu_{\alpha\beta} - \partial_\beta \Gamma^\nu_{\mu\alpha} + \Gamma^\ell_{\alpha\beta}\, \Gamma^\nu_{\mu\ell} - \Gamma^\ell_{\mu\alpha}\, \Gamma^\nu_{\beta\ell}} \tag{1.3}$$

$$\boxed{R_{\alpha\beta} \stackrel{\text{def}}{=} R^\mu{}_{\alpha\mu\beta} = \partial_\mu \Gamma^\mu_{\alpha\beta} - \partial_\beta \Gamma^\mu_{\alpha\mu} + \Gamma^\sigma_{\alpha\beta}\, \Gamma^\mu_{\sigma\mu} - \Gamma^\sigma_{\alpha\mu}\, \Gamma^\mu_{\sigma\beta}} \tag{1.4}$$

5. A parameterized curve $q^\alpha(\xi)$ can be interpreted as a motion with respect to the 'time' ξ of a point on the manifold Q. The vector

$$\boxed{v^\alpha(\xi) \stackrel{\text{def}}{=} \frac{dq^\alpha}{d\xi}} \tag{1.5}$$

[2] A definition or equation that is expressed by recurring to a coordinate system is said to have an **intrinsic meaning**, or even a **geometrical meaning**, if it does not depend on the choice of the coordinates.

[3] These definitions are a matter of conventions. They can change both for the position of the indices and for the sign.

can then be interpreted as **velocity** of the point. A connection allows to define the **acceleration vector**

$$\boxed{a^{\gamma}(\xi) \stackrel{\text{def}}{=} \frac{d^2 q^{\gamma}}{d\xi^2} + \Gamma^{\gamma}_{\alpha\beta} \frac{dq^{\alpha}}{d\xi} \frac{dq^{\beta}}{d\xi}} \tag{1.6}$$

Both of these definitions are independent of the choice of coordinates.

6. A curve is called **geodesic** if its acceleration is always parallel to the velocity,

$$a^{\gamma}(\xi) = \lambda(\xi) \frac{dq^{\gamma}}{d\xi}, \tag{1.7}$$

that is, if there exists a function $\lambda(\xi)$ for which the **equations of geodesics**

$$\boxed{\frac{d^2 q^{\gamma}}{d\xi^2} + \Gamma^{\gamma}_{\alpha\beta} \frac{dq^{\alpha}}{d\xi} \frac{dq^{\beta}}{d\xi} = \lambda(\xi) \frac{dq^{\gamma}}{d\xi}} \tag{1.8}$$

are fulfilled.

7. The parameter of a curve can be changed. Both the speed and the acceleration then change. A geodesic can admit an **affine parameter** defined up to an affine transformation, for which the acceleration vanishes, so that Eq. (1.8) become

$$\frac{d^2 q^{\gamma}}{d\bar{\xi}^2} + \Gamma^{\gamma}_{\alpha\beta} \frac{dq^{\alpha}}{d\bar{\xi}} \frac{dq^{\beta}}{d\bar{\xi}} = 0. \tag{1.9}$$

8. A manifold is said to be **Riemannian** if is endowed by a double symmetric tensor field $g_{\alpha\beta} = g_{\beta\alpha}$, non-degenerate $\det[g_{\alpha\ beta}] \neq 0$, called **metric tensor** or briefly **metric**. If its signature is positive the manifold is called **properly Riemannian** or simply **Riemannian**; otherwise it is called **semi-Riemannian** or **pseudo-Riemannian**. A metric tensor is used first of all to define the **scalar product** (also called **dot product**) between two vectors:

$$\boldsymbol{u} \cdot \boldsymbol{v} \stackrel{\text{def}}{=} g_{\alpha\beta}\, u^{\alpha}\, v^{\beta}.$$

9. Whatever the signature of the metric, a Riemannian manifold admits a unique 'canonical' connection, called **Levi-Civita connection**. This is a linear and symmetric connection whose transport preserves the scalar product. It can be shown that its symbols are given by

$$\Gamma^{\gamma}_{\alpha\beta} \stackrel{\text{def}}{=} \tfrac{1}{2}\, g^{\gamma\delta} \left(\partial_{\alpha} g_{\beta\delta} + \partial_{\beta} g_{\delta\alpha} - \partial_{\delta} g_{\alpha\beta}\right). \tag{1.10}$$

These are called **Christoffel symbols**, more precisely **Christoffel symbols of second kind**. Those of **first kind** are obtained by lowering the upper index by means of the metric tensor:

$$\Gamma_{\alpha\beta\,\delta} \stackrel{\text{def}}{=} \Gamma^{\gamma}_{\alpha\beta}\, g_{\gamma\delta} = \tfrac{1}{2}\left(\partial_\alpha g_{\beta\delta} + \partial_\beta g_{\delta\alpha} - \partial_\delta g_{\alpha\beta}\right). \tag{1.11}$$

10. Beside the Riemann and the Ricci tensors (1.3) and (1.4) for a Levi-Civita connection we can also define the **totally covariant Riemann tensor**

$$\boxed{R_{\lambda\alpha\mu\beta} \stackrel{\text{def}}{=} g_{\lambda\nu}\, R^{\nu}_{\ \alpha\mu\beta}} \tag{1.12}$$

and the **Ricci curvature** (or **Ricci scalar**)

$$\boxed{R \stackrel{\text{def}}{=} g^{\alpha\beta}\, R_{\alpha\beta}} \tag{1.13}$$

11. It can be shown that if equation

$$\boxed{R_{\alpha\beta\gamma\delta} = K\,(g_{\alpha\gamma}\, g_{\beta\delta} - g_{\alpha\delta}\, g_{\beta\gamma})} \tag{1.14}$$

is satisfied then the factor K is a constant, called **curvature constant**, and the Riemannian manifold, or the metric tensor, are said to have **constant curvature**. Equation (1.14) is equivalent to

$$R^{\alpha}_{\ \beta\gamma\delta} = K\,(\delta^{\alpha}_{\gamma}\, g_{\beta\delta} - \delta^{\alpha}_{\delta}\, g_{\beta\gamma}). \tag{1.15}$$

It follows that on a manifold of dimension n with constant curvature K the Ricci tensor and the Ricci scalar respectively take on the expressions

$$\boxed{R_{\alpha\beta} = (n-1)\,K\,g_{\alpha\beta}} \quad \boxed{R = n\,(n-1)\,K} \tag{1.16}$$

Remark 1.2 It is important to note that the Christoffel symbols (1.10) are invariant under **conformal transformations** of the metric

$$g_{\alpha\beta} \mapsto \widehat{g}_{\alpha\beta} = \sigma\, g_{\alpha\beta}$$

with constant factor $\sigma \in \mathbb{R}$. It follows that the components of the Riemann and Ricci tensors are also invariant:

$$\widehat{R}^{\alpha}_{\ \beta\gamma\delta} = R^{\alpha}_{\ \beta\gamma\delta}, \quad \widehat{R}_{\alpha\beta} = R_{\alpha\beta}. \quad \bullet$$

Remark 1.3 If a metric $g_{\alpha\beta}$ has constant curvature K then every metric conformal to it $\widehat{g}_{\alpha\beta} = \sigma\, g_{\alpha\beta}$ with constant factor σ has constant curvature given by

$$\boxed{\widehat{K} = \frac{K}{\sigma}} \tag{1.17}$$

In fact, if $g_{\alpha\beta}$ has constant curvature then the Eq. (1.15) is satisfied and therefore

$$\widehat{R}^{\alpha}{}_{\beta\gamma\delta} = K\,(\delta^{\alpha}_{\gamma}\,g_{\beta\delta} - \delta^{\alpha}_{\delta}\,g_{\beta\gamma}) = K\,\sigma^{-1}\,(\delta^{\alpha}_{\gamma}\,\widehat{g}_{\beta\delta} - \delta^{\alpha}_{\delta}\,\widehat{g}_{\beta\gamma}),$$

i.e.

$$\widehat{R}^{\alpha}{}_{\beta\gamma\delta} = \widehat{K}\,(\delta^{\alpha}_{\gamma}\,\widehat{g}_{\beta\delta} - \delta^{\alpha}_{\delta}\,\widehat{g}_{\beta\gamma}),$$

with $\widehat{K} \stackrel{\text{def}}{=} K/\sigma$. •

Remark 1.4 It can be shown that in a neighborhood of every point of a constant curvature manifold K there exist coordinates (x_i), called **curvature coordinates** or **Riemann coordinates**, such that ds^2 takes the **Riemann form**

$$\boxed{ds^2 = \frac{\sum_i e_i\,(dx_i)^2}{\left(1 + \frac{1}{4}\,K\,\sum_i e_i\,x_i^2\right)^2}, \quad e_i = \pm 1} \tag{1.18}$$

For a positive definite metric, for which $e_i = 1$, we have

$$\boxed{ds^2 = \frac{\sum_i (dx_i)^2}{\left(1 + \frac{1}{4}\,K\,\sum_i x_i^2\right)^2}} \tag{1.19}$$

The curvature coordinates are **orthogonal**: $g_{ij} = 0$ for $i \neq j$. Such a coordinate system is determined by a point p_0, where the coordinates vanish, and by a basis $\boldsymbol{u}_i^0$ of unit vectors applied in p_0 orthogonal to each other. The presence of such a point is not in conflict with the isotropy principle. A notable example of Riemann coordinates is provided by the stereographic projection of the sphere $\mathbb{S}_n$ (Sect. 7.1). •

1.5 Permanence of the Sign of Spatial Curvature

We denote by $K(t)$ or by K_t the curvature constant of a spatial section (S_t, g_t) (see Theorem 1.4). This 'constant' is a function of t over the whole time interval (t_α, t_ω) of the life of the Universe.

Theorem 1.5 (i) *If $K_{t_1} \neq 0$ then for every other time $t_2 \neq t_1$ thecurvatureK_{t_2} has the same sign as K_{t_1} and the two metrics g_{t_1} and g_{t_2} are related by the conformal transformation*

$$\boxed{g_{t_1} = \frac{K_{t_2}}{K_{t_1}}\,g_{t_2}} \tag{1.20}$$

(ii) *If $K_{t_1} = 0$ then $K_{t_2} = 0$ in ogni t_2.*

Proof Let us denote by R_t the Ricci tensor of the spatial metric g_t. From the first of (1.16) it follows that

$$R_{t_1} = 2\, K_{t_1}\, g_{t_1}, \quad R_{t_2} = 2\, K_{t_2}\, g_{t_2}.$$

Supposing $K_{t_1} \neq 0$ and setting

$$\sigma = \frac{K_{t_2}}{K_{t_1}},$$

then, as stated in Remark 1.3 and for the Eq. (1.17), the curvature of the conformal metric $\widehat{g} = \sigma\, g_{t_2}$ is

$$\widehat{K} = \frac{K_{t_2}}{\sigma} = K_{t_1}.$$

By virtue of (1.16) the Ricci tensor of $\widehat{g}$ is equal to

$$[*] \quad \widehat{R} = 2\, K_{t_1}\, g_{t_1}.$$

On the other hand, as stated in Remark 1.2 we have $\widehat{R}_{ab} = R_{t_2}$, therefore

$$[**] \quad \widehat{R} = R_{t_2} = 2\, K_{t_2}\, g_{t_2}.$$

From these two last equations $[*]$ and $[**]$ it follows that

$$\boxed{K_{t_1}\, g_{t_1 ab} = K_{t_2}\, g_{t_2 ab}} \tag{1.21}$$

and (1.20) is proved. Since both metrics g_{t_1} and g_{t_2} are positive definite, in (1.20) K_{t_1} and K_{t_2} must have the same sign. (ii) As demonstrated above, the two conditions $K_{t_1} = 0$ and $K_{t_2} \neq 0$ are contradictory. ■

Remark 1.5 Theorem 1.5 states that *the sign of the spatial curvature $K(t)$ does not change throughout the open time interval (t_α, t_ω) of the life of the Universe*. The sign of the curvature dictates the topological characteristics of the spatial sections S_t.[4] For example, for $K(t) = 0$ the spatial section S_t can be homeomorphic to the three-dimensional Euclidean space $\mathbb{E}_3$, or to a torus $\mathbb{T}_3$. Instead, for $K(t) > 0$ the S_t can be a three-dimensional sphere $\mathbb{S}_3$. In any case, even if $K(t)$ is a continuous function, a change in its sign would produce a discontinuity in the topology. Thanks to the Theorem 1.5 such an unacceptable singularity is avoided. •

---

[4] The famous **Killing-Hopf theorem**, see for instance [16, 22], is the basic tool for the topological classification of manifolds with constant curvature.

1.6 Conformal Factor Between Two Spatial Metrics

In the case of non-zero curvature, since the spatial curvatures at two different times K_{t_1} and K_{t_2} have the same sign, the ratio K_{t_2}/K_{t_1} is positive, so that we can introduce the following positive function in two time variables:

$$\boxed{a(t_1, t_2) \stackrel{\text{def}}{=} \sqrt{\frac{K_{t_2}}{K_{t_1}}}} \tag{1.22}$$

Consequently Eq. (1.20) takes the form

$$\boxed{g_{t_1} = a^2(t_1, t_2)\, g_{t_2}} \tag{1.23}$$

and we can therefore state that (see Fig. 1.3)

Theorem 1.6 *There exists a positive function $a(t_1, t_2)$, with $t_1, t_2 \in (t_\alpha, t_\beta)$, such that two spatial metrics are related by the conformal transformation* (1.23).

Remark 1.6 In the case of zero curvature the existence of a function $a(t_1, t_2) > 0$ satisfying Eq. (1.23) follows from the general property that two Riemannian manifolds with the same dimension, the same signature and same curvature constant are locally isomorphic (see [7, 8]). •

Remark 1.7 Equation (1.23) is equivalent to

$$\boxed{ds_{t_1} = a(t_1, t_2)\, ds_{t_2}} \tag{1.24}$$

where ds_{t_1} and ds_{t_2} are the arc-elements of the metrics g_{t_1} e g_{t_2}. •

Remark 1.8 From Eq. (1.23) we derive the following properties of the conformal factor $a(t_1, t_2)$:

$$\boxed{\begin{array}{ll} a(t,t) = 1, & \text{normalization,} \\ a(t_1, t_2)\, a(t_2, t_3) = a(t_1, t_3) & \text{composition,} \\ a(t_2, t_1) = \dfrac{1}{a(t_1, t_2)} & \text{inversion.} \end{array}} \tag{1.25}$$

They will be applied frequently without explicit mention. •

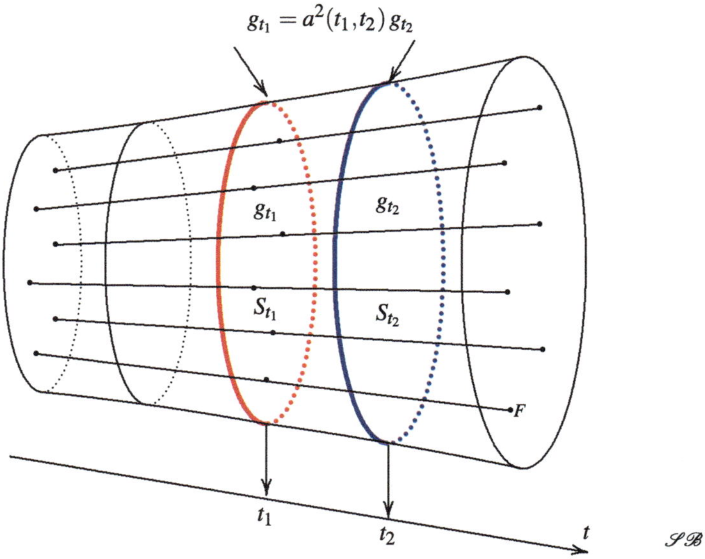

Fig. 1.3 Conformal transformation (1.23) between two spatial metrics

1.7 Reference Time and Scale Factor

By imposing a value $t_\sharp$ to t_2 and leaving $t_1 = t$ free to vary throughout the interval (t_α, t_ω) we obtain a function $a(t, t_\sharp)$ of the single variable t which we call **scale factor with reference time** $t_\sharp$. We call **reference space** the spatial section $S_{t_\sharp}$ equipped with the **reference metric** $g_{t_\sharp}$. For $t = t_\sharp$ we have $a(t_\sharp, t_\sharp) = 1$. The reference time can then be interpreted as **normalization time** for the scale factor. From (1.23) we derive the equation

$$\boxed{g_t = a^2(t, t_\sharp)\, g_{t_\sharp}} \tag{1.26}$$

which in the following will often be referred to as **factorization relation** of spatial metrics (Fig. 1.4).

Remark 1.9 The scale factor $a(t, t_\sharp)$ never vanishes in the open interval (t_α, t_ω) because the spatial metrics are always regular. •

By applying the rules (1.25) it can be shown that the following link exists between the scale factors referring to two different times $a(t, t_\sharp)$ and $a(t, t_\flat)$

$$\boxed{a(t, t_\sharp) = a(t, t_\flat)\, a(t_\flat, t_\sharp)} \tag{1.27}$$

Therefore they differ by the constant factor $a(t_\flat, t_\sharp)$ depending on the two reference times. Notice that this link looks like a *chain rule*.

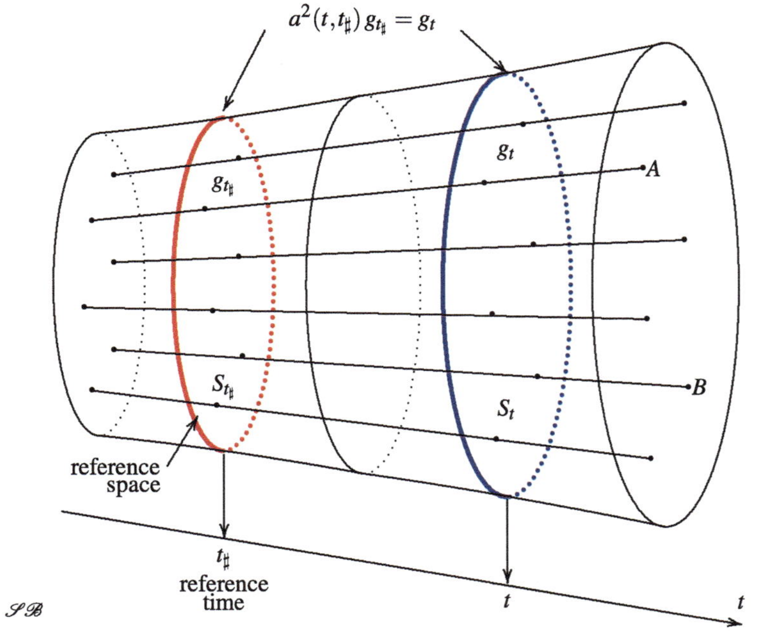

Fig. 1.4 Scale factor $a(t, t_\sharp)$ with reference time $t_\sharp$

From (1.22) and (1.24) we also derive the equations

$$\boxed{K(t) = \frac{K_\sharp}{a^2(t, t_\sharp)}, \quad K_\sharp \overset{\text{def}}{=} K(t_\sharp)} \tag{1.28}$$

$$\boxed{ds_t = a(t, t_\sharp)\, ds_\sharp, \quad ds_\sharp \overset{\text{def}}{=} ds(t_\sharp)} \tag{1.29}$$

which will be applied several times in the following.

Remark 1.10 The scale factor is a dimensionless function of cosmic time t that, as will be seen, holds all the information about the time evolution of the Universe and most cosmological quantities. The fact that in our axiomatic approach the scale factor $a(t, t_\sharp)$ turns out to be dependent on the choice of a reference time represents a novelty with respect to the current literature. This dependence is not only of theoretical interest but also provides a solving in checking the correctness of statements, definitions and equations. It is indeed clear that, to make geometric or physical sense, statements, equations and definitions involving it, possibly together with its derivatives, must be independent of the choice of $t_\sharp$. An example is the definition (1.33) of the Hubble factor in the next section. •

1.8 Recession Speed and Hubble Law

Two types of distances are defined between two cosmic bodies A and B (see Fig. 1.5 below):

• The **synchronous distance** $d_{AB}(t)$ at time t is the distance measured in the metric g_t of the spatial section S_t, i.e. the length of the geodesic that joins the intersection points of the histories of A and B with S_t.

• The **reference distance** $d_{AB}(t_\sharp)$, also called **co-moving distance**, is the distance measured in the reference space $(S_{t_\sharp}, g_{t_\sharp})$.

By virtue of Eq. (1.29) the relationship

$$d_{AB}(t) = a(t, t_\sharp)\, d_{AB}(t_\sharp) \tag{1.30}$$

holds between these two distances. Since $t_\sharp$ is fixed, differentiating this equation with respect to t we find

$$\dot{d}_{AB}(t) = \dot{a}(t, t_\sharp)\, d_{AB}(t_\sharp). \tag{1.31}$$

In turn, by applying (1.30), we get

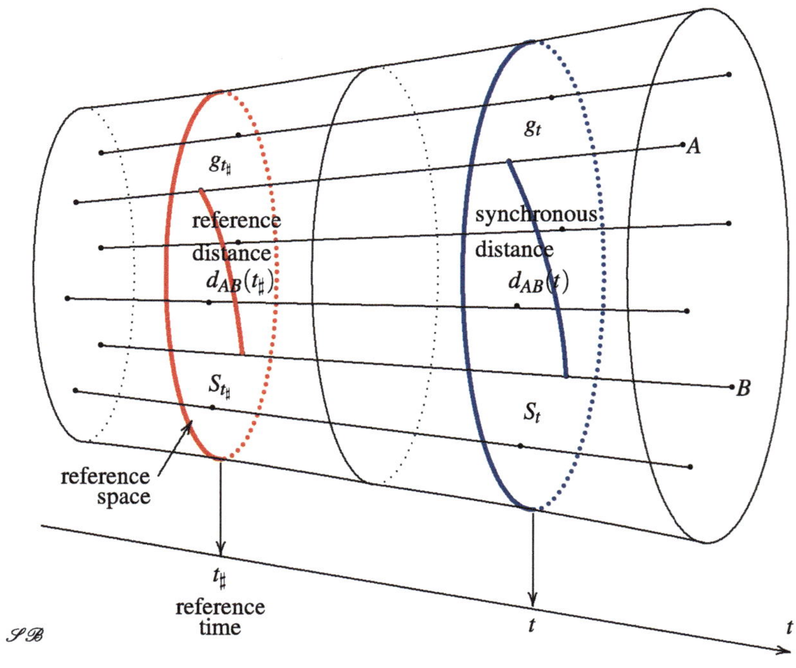

Fig. 1.5 Distances between two galaxies A and B

$$\boxed{\dot{d}_{AB}(t) = \frac{\dot{a}(t, t_\sharp)}{a(t, t_\sharp)} d_{AB}(t)} \tag{1.32}$$

This equation holds whatever reference time $t_\sharp$, so we can introduce the function of the single variable t

$$\boxed{H(t) \stackrel{\text{def}}{=} \frac{\dot{a}(t, t_\sharp)}{a(t, t_\sharp)}} \tag{1.33}$$

called **Hubble factor** or **Hubble parameter**, and rewrite (1.32) in the form

$$\boxed{\dot{d}_{AB}(t) = H(t)\, d_{AB}(t)} \tag{1.34}$$

This is the well-known **Hubble law**. The derivative $\dot{d}_{AB}(t)$ is called **recession speed** of the galaxies A and B at time t.

Remark 1.11 Since $a(t_\sharp, t_\sharp) = 1$ (normalization property) the value of the Hubble factor (1.33) at the reference time $t_\sharp$ is

$$\boxed{H(t_\sharp) = \dot{a}(t_\sharp, t_\sharp)} \tag{1.35}$$

This formula will be useful later on. •

1.9 Dimensional Analysis

An equation cannot be considered correct if it is not consistent with the physical dimension of the quantities involved. Unfortunately this **principle of dimensional homogeneity** is not always respected. This causes reprehensible misunderstandings. Let's look at three examples.

(i) A widespread habit is to set c = speed of light = 1. With this action the symbol c disappears from the formulas and their dimensional coherence is lost. The same happens for other physical constants, such as Planck's constant. Of course it is completely legitimate to set $c = 1$ if it is considered convenient for the calculations. This only involves a change in the units of measurement of lengths and times. But if you opt for this choice it is extremely important *to keep the symbol c in the formulas* even if its numerical value is 1.

(ii) On a Riemannian manifold one can consider coordinates of different dimensions: angles (dimensionless), times, lengths, etc. As a result, the various components of the metric tensor can have different dimensions. This difference has strong repercussions on the components of the curvature tensors, therefore, in particular, on Einstein's equations. As will be seen in the next section, to avoid this confusion it will be convenient to use on cosmic space-time only length-dimensional coordinates, including *time*.

(iii) The exponential function e^z, where z is a real or complex number, is defined by a power series of z which, if it represents a physical or geometric quantity, must be dimensionless, otherwise it would make no sense to add z to z^2, etc. The same consideration must be given to the extension of e^z to the case where z is a square matrix: its elements must be dimensionless. There are many examples in the literature where due attention is not paid in this regard.

In the following we will use the symbol Dim (X) to indicate the dimension of a physical-geometrical entity X.[5] The fundamental dimensions are denoted as follows:

$$\begin{cases} \text{Dim (time)} = \mathsf{T} \\ \text{Dim (length)} = \mathsf{L} \\ \text{Dim (mass)} = \mathsf{M} \\ \text{Dim (dimensionless quantity)} = 1 \end{cases}$$

The physical dimension of a quantity X can be expressed as the product of positive or negative integer powers of the symbols T, L and M,

$$\text{Dim}\,(X) = \mathsf{T}^a\,\mathsf{L}^b\,\mathsf{M}^c, \quad a, b, c \in \mathbb{Z}.$$

The dimensions of the recurring fundamental quantities are listed in the following tables (Tables 1.1, 1.2 and 1.3).

Table 1.1 Geometrical and kinematical quantities

Quantity	Dim	Quantity	Dim
Area	L^2	Acceleration	$\mathsf{L}\,\mathsf{T}^{-2}$
Volume	L^3	Angle	1 (dimensionless)
Velocity	$\mathsf{L}\,\mathsf{T}^{-1}$	Angular velocity	T^{-1}

Table 1.2 Physical quantities

Quantity	Dim
Force (or mass times acceleration)	$\mathsf{M}\,\mathsf{L}\,\mathsf{T}^{-2}$
Pression (force/area)	$\mathsf{M}\,\mathsf{L}^{-1}\,\mathsf{T}^{-2}$
Energy, work (force times length)	$\mathsf{M}\,\mathsf{L}^{2}\,\mathsf{T}^{-2}$
Energy density (energy/volume)	$\mathsf{M}\,\mathsf{L}^{-1}\,\mathsf{T}^{-2}$
Mass density (mass/volume)	$\mathsf{M}\,\mathsf{L}^{-3}$

[5] The symbol $[X]$ is also widely used.

Table 1.3 Scale and Hubble factors

Quantity	Dim
Scale factor $a(t)$	1 (dimensionless)
Hubble factor $H(t)$	T^{-1}

1.10 Co-moving Coordinates

Since the reference space $S_{t_\sharp}$ is diffeomorphic to all other spatial sections, if we take a coordinate system $(q^a) = (q^1, q^2, q^3)$ on an open domain $U \subseteq S_{t_\sharp}$ then we generate a coordinate system (t, q^a) on the open tubular sub-set of M made of the galactic histories that intersect U (Fig. 1.6).

Coordinates of this type are called **co-moving coordinates**. The three spatial coordinates (q^a) are **Lagrangian coordinates** of the cosmic fluid: they are constant along the histories crossing U.

In the following we will denote with g_{tab} and $g_{\sharp ab}$ the components with respect to the coordinates (q^a) of the spatial metric tensors of S_t and $S_{t_\sharp}$ respectively. They satisfy the conformal relation (1.26):

$$\boxed{g_{tab} = a^2(t, t_\sharp)\, g_{\sharp ab}} \tag{1.36}$$

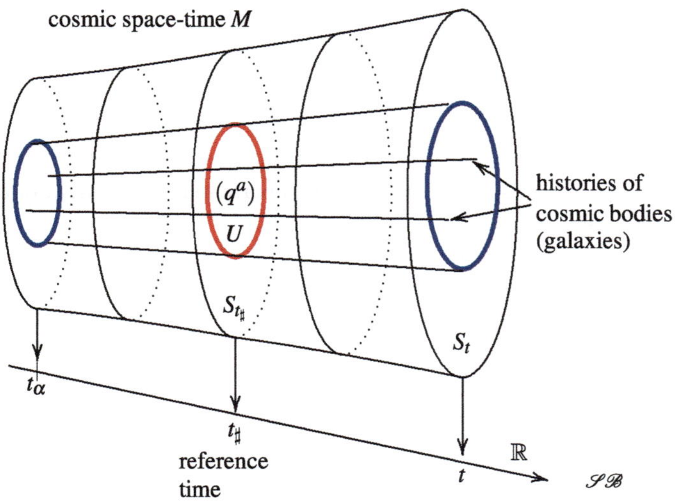

Fig. 1.6 Co-moving coordinates

We will not use particular types of spatial coordinates (q^a). We will only impose the condition that they are length-dimensional, so that the components of the metric tensor g_{tab} are dimensionless. Consequently, for reasons of homogeneity, we will replace the time coordinate t with a L-dimensional coordinate q^0 through the simple linear relation

$$\boxed{q^0 = \kappa\, t} \tag{1.37}$$

where κ is an arbitrarily fixed positive constant having the dimension of a velocity. We call it **auxiliary velocity**.[6]

In the following we will always and tacitly refer to coordinates $(q^\alpha) = (q^0, q^a)$ of this type. The spatial coordinates (q^a) will remain generic.

1.11 Isotropic Vectors and Tensors

We say that a vector field or a tensor field in the cosmic space-time M is **isotropic** if it does not induce particular vector fields on the spatial sections. Hence, in an isotropic cosmological model non-isotropic vector or tensor fields **are not admissible**.

Theorem 1.7 *A vector field V^α is isotropic if and only if its components with respect to a co-moving coordinate system $(q^\alpha) = (q^0, q^a)$ are of the type*

$$\begin{cases} V^0 = \text{function of } q^0 \text{ only}, \\ V^a = 0. \end{cases} \tag{1.38}$$

Proof (i) If the function V^0 also depends on the spatial coordinates (q^a) then on each spatial section its gradient would define a particular vector field in contrast to the isotropy principle.

(ii) For the co-moving coordinates $(q^\alpha) = (q^0, q^a)$ only transformations of the spatial coordinates (q^a) are admissible since the coordinate $q^0 = \kappa\, t$ is uniquely defined.[7] Therefore the spatial components V^a define a particular spatial vector in contrast to the isotropy principle. Hence they must be zero. ∎

Theorem 1.8 *A contravariant double symmetric tensor field is isotropic if and only if its components $T^{\alpha\beta}$ with respect to a co-moving coordinate system $(q^\alpha) = (q^0, q^a)$ are of the type*

[6] When we will deal with relativistic cosmology we will be led to consider $\kappa = c$.

[7] This must also be taken into account in the following.

$$\begin{cases} T^{00} = \phi(q^0) = \text{function of } q^0 \text{ only,} \\ T^{0a} = 0, \\ T^{ab} = \psi(q^0)\, g_\sharp^{ab}(\widetilde{q}) = \text{a function of } q^0 \text{ times } g_\sharp^{ab}, \end{cases} \tag{1.39}$$

where $g_\sharp^{ab}(\widetilde{q})$ are the contravariant components of the reference metric, having denoted by $\widetilde{q}$ any spatial coordinate system (q^a).

Proof Transformation law of the components of a contravariant tensor:

$$T^{\alpha\beta} = J^{\alpha}_{\alpha'}\, J^{\beta}_{\beta'}\, T^{\alpha'\beta'}, \quad J^{\alpha'}_{\alpha} \overset{\text{def}}{=} \frac{\partial q^{\alpha'}}{\partial q^{\alpha}}, \quad J^{\alpha}_{\alpha'} \overset{\text{def}}{=} \frac{\partial q^{\alpha}}{\partial q^{\alpha'}}.$$

For a co-moving coordinate transformation that leaves q^0 unchanged we have

$$J^{0'}_{0} = 1, \quad J^{0'}_{a} = 0, \quad J^{a'}_{0} = 0, \quad J^{0}_{c'} = 0, \quad J^{c}_{0'} = 0.$$

As a consequence,

$$\begin{bmatrix} T^{00} = J^{0}_{\alpha'}\, J^{0}_{\beta'}\, T^{\alpha'\beta'} = (J^{0}_{0'})^2\, T^{0'0'} = T^{0'0'}. \\ T^{0b} = J^{0}_{\alpha'}\, J^{b}_{\beta'}\, T^{\alpha'\beta'} = J^{0}_{0'}\, J^{b}_{b'}\, T^{0'b'} = J^{b}_{b'}\, T^{0'b'}. \\ T^{ab} = J^{a}_{\alpha'}\, J^{b}_{\beta'}\, T^{\alpha'\beta'} = J^{a}_{a'}\, J^{b}_{b'}\, T^{a'b'}. \end{bmatrix}$$

These equations show that: (i) T^{00} is a scalar field, so it must be a function of only q^0; (ii) T^{0b} is a space vector field, so it must be zero; (iii) T^{ab} is a symmetric tensor on every spatial section therefore it should not generate particular eigenvectors; it can only be proportional to the metric with a coefficient depending at most on q^0. ■

Remark 1.12 A similar result holds for a covariant symmetric two-tensor field $T_{\alpha\beta}$:

$$\begin{cases} T_{00} = \phi(q^0) = \text{function of } q^0 \text{ only ,} \\ T_{0a} = 0, \\ T_{ab} = \psi(q^0)\, g_{\sharp ab}(\widetilde{q}) = \text{function } q^0 \text{ times } g_{\sharp ab} \end{cases} \tag{1.40}$$

Remark 1.13 This theorem shows that every isotropic symmetric two-tensor field, whether contravariant or covariant, is completely determined by two functions $\phi(q^0)$ and $\psi(q^0)$ of q^0 which we call **characteristic functions** of the tensor. •

Remark 1.14 An anti-symmetric double contravariant tensor $F^{\alpha\beta}$ gives rise to a spatial vector F^{0a} and a spatial anti-symmetric tensor F^{ab}. For isotropy it must be $F^{0a} = 0$. However, every anti-symmetric two-tensor F^{ab} on a three-dimensional Riemannian manifold admits real eigenvectors. This goes against the isotropy principle. Consequently, an isotropic anti-symmetric two-tensor is necessarily zero. This is for example the case of the electromagnetic tensor. Therefore an electromagnetic field should not appear in an isotropic cosmological model, for example in Einstein's equations. However, the presence of a large number of electromagnetic fields produces a large number of spatial eigenvectors such as to make the isotropy principle effectively respected, so it can be summarized in an isotropic scalar field (i.e. dependent on q^0 only) as a indexradiation density **radiation density**. •

1.12 Cosmic Monitor and Wandering Particles

Just as for any other spatial section, the reference space is in one-to-one correspondence with the galactic histories that cross it transversally and therefore with the set of all galaxies. Suppose that there is a control device for this set, say a **cosmic monitor**, made up of points (pixels), each of which represents a galaxy.

Suppose further that this monitor is under the control of an **Astronomer**. The monitor and the Astronomer are drawn at the bottom of Fig. 1.7. Just for fun the Astronomer wears the clothes and hat of a wizard. The M spacetime is represented at the top, along with the galactic curves and space slices. For simplicity the reference space in denoted by $\widetilde{S}$.

Remark 1.15 Notation to keep in mind. In the following all the entities that are in the reference space, and that are perceived by the cosmic monitor, will be marked by a superimposed tilde $\widetilde{\ }$:

$\widetilde{q}$, any spatial coordinate system (q^a);

$\widetilde{g}, \widetilde{g}_{ab}$, reference metric and its covariant components;

$\widetilde{\Gamma}^c_{ab}$, Christoffel symbols of the metric $\widetilde{g}$; etc. •

The cosmic monitor is a snapshot of the cosmos at time $t_\sharp$ and is therefore a sort of three-dimensional 'crystal ball' for the magician who observes it. In this sphere the world is frozen at time $t_\sharp$. The galaxies are stationary and no expansion or contraction is felt.[8] The Astronomer has a clock that indicates the cosmic time t,[9] but he does not see the space-time, which in the figure is above his head. However, even though the galaxies remain fixed, he observes on the monitor the trace of moving points which he calls **wandering particle**.

For us, who are aware about the existence of cosmic space-time, the traces that these particles leave on the monitor are the projections of histories of tiny objects wandering in the Universe. They could be spacecrafts on intergalactic travel, or comets, or physical particles, such as photons.

[8] If we like, it could be interpreted as the **firmament** of biblical cosmology.

[9] He is a very long-lived and patient astronomer, his times are of the order of millions of years.

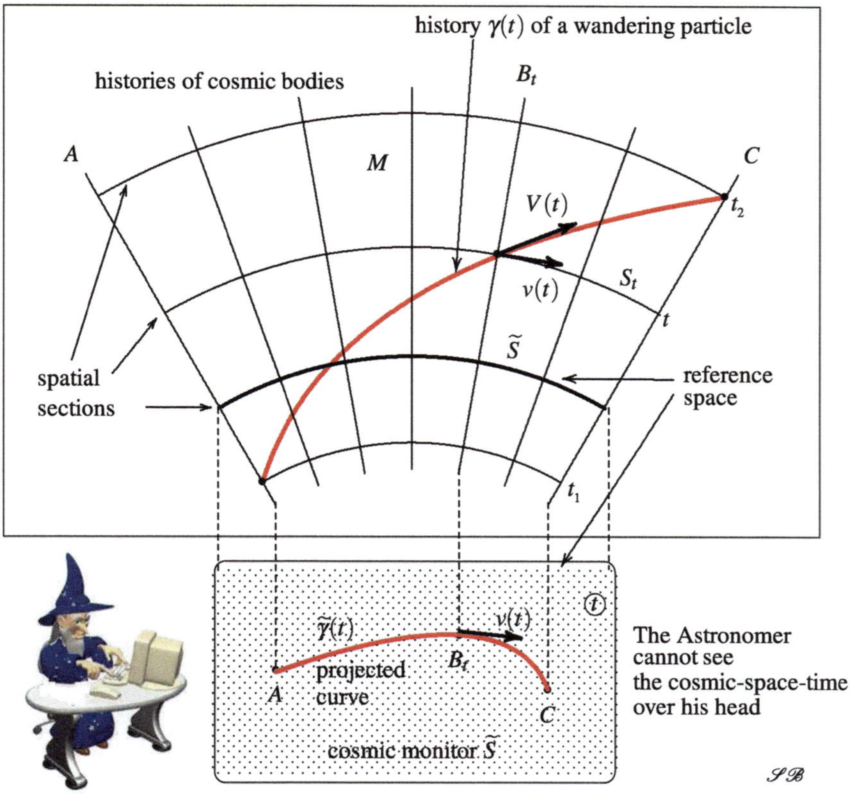

Fig. 1.7 The Astronomer observes the trace of a wandering particle on the cosmic monitor

In Fig. 1.7 we can see the $\gamma(t)$ and $\widetilde{\gamma}(t)$ curves of a wandering particle starting from a galaxy A at time t_1, which reaches another galaxy C at time t_2 and which crossed a galaxy B at an intermediate time t.

However, the Astronomer notes that some of the observed traces are geodesics (Fig. 1.8). He can distinguish them because, let's remember, the monitor has a metric, which is identified with that of the reference space, and furthermore, being also a Mathematician, the Astronomer is acquainted with this notion.

Thus the Astronomer perceives the idea that in the Universe there are **special wandering particles** of which, however, he does not know the physical nature. As a mathematician he then conjectures that the space-time curves from which these traces originate are also geodesics. But, he observes, for this conjecture to make sense, the cosmic space-time must be endowed with some particular connection.

Let us make this observation our own.

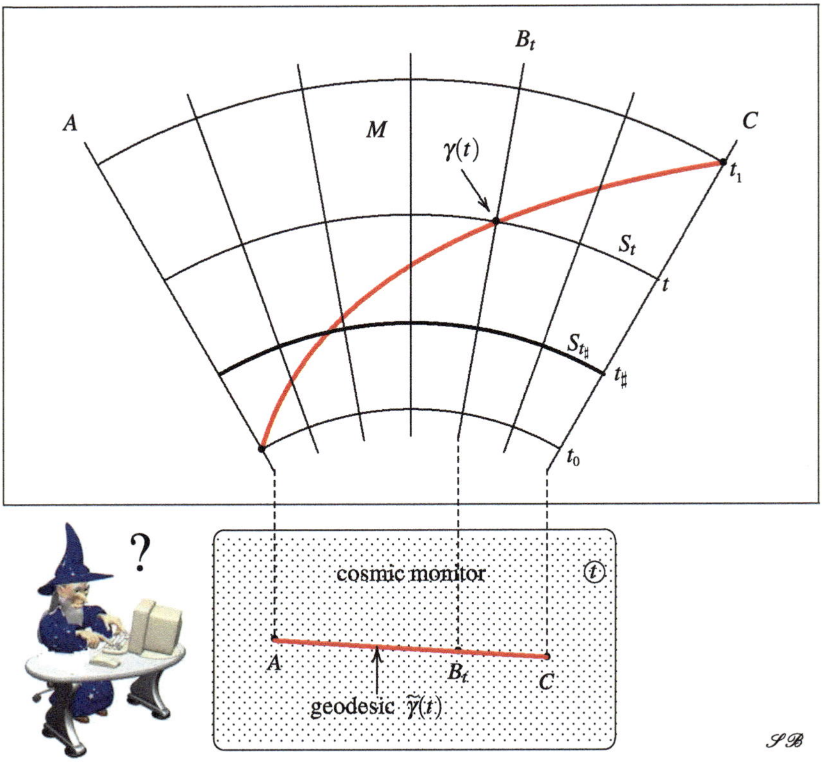

Fig. 1.8 The Astronomer observes, rather surprised, the geodesic trace of a 'special' particle

In the next section we will investigate about the existence of a connection in some way compatible with the various structures introduced so far into cosmic space-time by virtue of geometrical postulates. It should be noted at once that this will not be a Levi-Civita connection because in space-time there does not exist, at least for now, a four-dimensional metric tensor. Metric tensors are present in the spatial sections only.

1.13 Cosmic Connections

In this section we investigate about the existence in cosmic space-time of linear and symmetric connections satisfying suitable **compatibility requirements** with the geometric structures internal to space-time known so far. We'll call them **cosmic connections**. The order in which the requirements are exposed is not reversible.

Requirement 1. *A cosmic connection $\Gamma = (\Gamma^{\gamma}_{\alpha\beta})$ must be isotropic, that is, it must not give rise to particular vector fields on the spatial sections.*

Theorem 1.9 *A connection is isotropic if and only if its symbols $\Gamma^{\gamma}_{\alpha\beta}$ in co-moving coordinates (q^0, q^a) are of the type*

$$\boxed{\begin{array}{l} \Gamma^{0}_{a0} = 0, \quad \Gamma^{c}_{00} = 0 \\ \Gamma^{c}_{a0} = E(q^0)\,\delta^{c}_{a}, \quad \Gamma^{0}_{ab} = F(q^0)\,\widetilde{g}_{ab}(\widetilde{q}), \quad \Gamma^{0}_{00} = G(q^0) \end{array}} \tag{1.41}$$

where E, F and G are functions of q^0 only and where furthermore the symbols with only Latin indices Γ^{c}_{ab} are symbols of a connection on each spatial section.

Trace of the proof. Let's recall the transformation law (1.2) of the connection symbols and apply it to co-moving coordinate transformations that leave q^0 unchanged,

$$q^{0'} = q^0, \qquad q^{a'} = q^{a'}(q^a)$$

and for which we have

$$J^{0'}_{0} = 1, \quad J^{0'}_{a} = 0, \quad J^{a'}_{0} = 0, \quad J^{0}_{c'} = 0, \quad J^{c}_{0'} = 0.$$

We then examine all the particular cases of (1.2). Just one example for brevity:

$$\Gamma^{0}_{a0} = J^{0}_{\gamma'}\,\Gamma^{\gamma'}_{\alpha'\beta'}\,J^{\alpha'}_{a}\,J^{\beta'}_{0} + J^{0}_{\gamma'}\partial_a J^{\gamma'}_{0} = J^{0}_{0'}\,\Gamma^{0'}_{a'0'}\,J^{a'}_{a}\,J^{0'}_{0} + J^{0}_{c'}\partial_a J^{c'}_{0} = \Gamma^{0'}_{a'0'}\,J^{a'}_{a}.$$

This result shows that the Γ^{0}_{a0} are the components of a particular differential 1-form on each space section. For the isotropy requirement it must vanish: $\Gamma^{0}_{a0} = 0$. ∎

With the symbols (1.41) the transport Eq. (1.1) and the geodesic Eq. (1.8) of an isotropic connection become

$$\begin{cases} \dfrac{dv^0}{d\xi} + G\,v^0\,\dfrac{dq^0}{d\xi} + F\,\widetilde{g}_{ab}\,v^a\,\dfrac{dq^b}{d\xi} = 0. \\[2ex] \dfrac{dv^c}{d\xi} + E\left(v^c\,\dfrac{dq^0}{d\xi} + v^0\,\dfrac{dq^c}{d\xi}\right) + \Gamma^{c}_{ab}\,v^a\,\dfrac{dq^b}{d\xi} = 0. \end{cases} \tag{1.42}$$

$$\begin{cases} \dfrac{d^2q^0}{d\xi^2} + F\,\widetilde{g}_{ab}\,\dfrac{dq^a}{d\xi}\,\dfrac{dq^b}{d\xi} + G\left(\dfrac{dq^0}{d\xi}\right)^2 = \lambda\,\dfrac{dq^0}{d\xi}, \\[2ex] \dfrac{d^2q^c}{d\xi^2} + \Gamma^{c}_{ab}\,\dfrac{dq^a}{d\xi}\,\dfrac{dq^b}{d\xi} + 2\,E\,\dfrac{dq^c}{d\xi}\,\dfrac{dq^0}{d\xi} = \lambda\,\dfrac{dq^c}{d\xi}. \end{cases} \tag{1.43}$$

For curves transversal to spatial sections the coordinate q^0 can be taken as a parameter. In this case $q^0(q^0) = q^0$ and $dq^0/dq^0 = 1$ and the previous equations become respectively

$$\begin{cases} \dfrac{dv^0}{dq^0} + G\, v^0 + F\, \tilde{g}_{ab}\, v^a\, \dfrac{dq^b}{dq^0} = 0. \\ \dfrac{dv^c}{dq^0} + E\left(v^c + v^0\, \dfrac{dq^c}{dq^0}\right) + \Gamma^c_{ab}\, v^a\, \dfrac{dq^b}{dq^0} = 0. \end{cases} \tag{1.44}$$

$$\begin{cases} F\, \tilde{g}_{ab}\, \dfrac{dq^a}{dq^0}\, \dfrac{dq^b}{dq^0} + G = \lambda, \\ \dfrac{d}{dq^0}\dfrac{dq^c}{dq^0} + \Gamma^c_{ab}\, \dfrac{dq^a}{dq^0}\, \dfrac{dq^b}{dq^0} + 2\, E\, \dfrac{dq^c}{dq^0} = \lambda\, \dfrac{dq^c}{dq^0}. \end{cases} \tag{1.45}$$

Theorem 1.10 *The histories of the galactic fluid are geodesics with respect to any isotropic connection.*

Proof The histories of the galactic fluid are transversal to the spatial sections for which the Eqs. (1.44) and (1.45) hold. Furthermore they are characterized by the equations q^a = constant so that (1.44) simplify to

$$\text{transport equations:} \begin{cases} \dfrac{dv^0}{dq^0} + G\, v^0 = 0, \\ \dfrac{dv^c}{dq^0} + E\, v^c = 0. \end{cases} \tag{1.46}$$

The equations of the second group (1.45) are identically satisfied, while the first equation reduces to

$$G = \lambda. \tag{1.47}$$

This equality provides the multiplier that satisfies the geodesic equations for the histories of the galactic fluid. ∎

Remark 1.16 The institution of a cosmic connection in cosmic space-time allows us to define the important notion of **free particle**: **it is a wandering particle whose history is a geodesic in space-time**. From Theorem 1.10 it follows that *the histories of the galactic fluid are free particles in <u>any</u> isotropic connection*. With the exception of this case, a free particle (‘*free-falling particle*’) should be understood as a particle that is passively subject to the action of the cosmic fluid. •

Requirement 2. *The coordinate q^0 is an affine parameter for galactic histories.*

Theorem 1.11 *Requirement 2 is satisfied if and only if $G = 0$.*

Proof This follows from (1.47). ∎

At this point the symbol table (1.41) becomes

$$\boxed{\begin{array}{l} \Gamma^0_{a0} = 0, \quad \Gamma^c_{00} = 0, \quad \Gamma^0_{00} = 0 \\ \Gamma^c_{a0} = E(q^0)\,\delta^c_a, \quad \Gamma^0_{ab} = F(q^0)\,\widetilde{g}_{ab}(\widetilde{q}) \\ \Gamma^c_{ab} \text{ are symbols of a spatial connection} \end{array}} \tag{1.48}$$

while the first transport Eq. (1.46) simply reduces to

$$v^0 = \text{constant}. \tag{1.49}$$

Theorem 1.12 *The Γ-transport along galactic histories brings space vectors to space vectors.*

Proof Space vectors are characterized by $v^0 = 0$ the transport Eq. (1.49) is satisfied. ■

By virtue of this theorem the following requirement makes sense

Requirement 3. *The dot product between space vectors is conserved by the transport along galactic histories.*

Remark 1.17 Note that the dot product is not required to be conserved in transport along any geodesic, nor along any curve, but along galactic histories. •

Now we see how the Hubble factor is involved.

Theorem 1.13 *Requirement 3 implies that the function $E(q^0)$ coincides with the Hubble factor thought of as a function of q^0, $E(q^0) = H(q^0)$.*

Proof The dot product of two space vectors $\boldsymbol{u} = [u^a(q^0)]$ and $\boldsymbol{v} = [v^a(q^0)]$ along a curve $q^\alpha(q^0)$ is given by

$$\boldsymbol{u}(q^0) \cdot \boldsymbol{v}(q^0) \stackrel{\text{def}}{=} g_{ab}(q^0, \widetilde{q})\, u^a(q^0)\, v^b(q^0) = a^2(q^0)\, \widetilde{g}_{ab}(\widetilde{q})\, u^a(q^0)\, v^b(q^0). \tag{1.50}$$

It follows that

$$\frac{d}{dq^0}(\boldsymbol{u} \cdot \boldsymbol{v}) = \widetilde{g}_{ab}(\widetilde{q}) \left[2\,a\,a'\,u^a\,v^b + a^2 \left(\frac{du^a}{dq^0}\,v^b + u^a\,\frac{dv^b}{dq^0} \right) \right].$$

Requirement 2 translates into equation $\dfrac{d}{dq^0}(\boldsymbol{u} \cdot \boldsymbol{v}) = 0$ which develops into

$$[*] \quad \widetilde{g}_{ab} \left[2\,a'\,u^a\,v^b + a \left(\frac{du^a}{dq^0}\,v^b + u^a\,\frac{dv^b}{dq^0} \right) \right] = 0, \quad \forall\, u^a, v^b.$$

Let us recall the second transport Eq. (1.46) of the space vectors, *so far not used*,

$$\frac{dv^c}{dq^0} + E\, v^c = 0.$$

Then

$$\left[\begin{array}{l} [*] \iff \widetilde{g}_{ab}\left[2\,a'\,u^a\,v^b - a\,E\left(u^a\,v^b + u^a\,v^b\right)\right] = 0 \\ \iff \widetilde{g}_{ab}\,u^a\,v^b\left[a' - a\,E\right] = 0, \quad \forall\, u^a, v^b \iff E = \dfrac{a'}{a} = H(q^0). \end{array}\right. \quad \blacksquare$$

After this theorem the symbol table (1.48) updates to the following:

$$\boxed{\begin{array}{l} \Gamma^0_{a0} = 0, \quad \Gamma^c_{00} = 0, \quad \Gamma^0_{00} = 0 \\ \Gamma^c_{a0} = H(q^0)\,\delta^c_a, \quad \Gamma^0_{ab} = F(q^0)\,\widetilde{g}_{ab}(\widetilde{q}) \\ \Gamma^c_{ab} \text{ are symbols of a spatial connection} \end{array}} \tag{1.51}$$

Remark 1.18 As we know the scale factor $a(t)$ can be seen as a function of the parameter $q^0 = \kappa\, t$ having the dimension of a length. Its derivative with respect to q^0 will be denoted by $a'(q^0)$. As regards the Hubble factor, passing from the parameter t to the parameter q^0, it results

$$H(t) \stackrel{\text{def}}{=} \frac{\dot{a}}{a} = a^{-1}\,\frac{da}{dq^0}\,\frac{dq^0}{dt} = \kappa\,\frac{a'}{a}.$$

Then if we set

$$H(q^0) \stackrel{\text{def}}{=} \frac{a'}{a} \tag{1.52}$$

we get

$$H(t) = \kappa\, H(q^0). \quad \bullet \tag{1.53}$$

Requirement 4 (the last one). *The Γ-geodesics transversal to the spatial sections are projected into geodesics of the reference space.*

Remark 1.19 This is the requirement that gave rise to the search for connections in space-time (end of Sect. 1.12). Note that this requirement is satisfied by galactic histories whose projections reduce to points of the reference space (i.e. to points on the monitor). 'Points' are in fact **singular geodesics**.[10] •

Theorem 1.14 *A curve $\gamma(q^0)$ transversal to the spatial sections is projected into a geodesic $\widetilde{\gamma}(q^0)$ of the reference space if and only if*

---

[10] The geodesic equations of a connection (1.8)

$$\frac{d^2q^\gamma}{d\xi^2} + \Gamma^\gamma_{\alpha\beta}\,\frac{dq^\alpha}{d\xi}\,\frac{dq^\beta}{d\xi} = \lambda(\xi)\,\frac{dq^\gamma}{d\xi}$$

are satisfied by the parametric curves $q^\alpha(\xi) =$ constant, which simply represent points.

$$\text{(i)}\quad \Gamma^c_{ab} = \widetilde{\Gamma}^c_{ab}$$

and

$$\text{(ii)}\quad \textit{by setting}\quad \boxed{V(q^0) \stackrel{\text{def}}{=} \frac{d\widetilde{s}}{dq^0} > 0}$$

where $d\widetilde{s}$ is the arc-element of the reference metric $\widetilde{g}$, equation

$$\boxed{\frac{d\log V}{dq^0} + 2\,H = F\,V^2} \tag{1.54}$$

holds.

Proof For curves transversal to spatial sections the coordinate q^0 can be taken as a parameter. In this case $q^0(q^0) = q^0$ and $dq^0/dq^0 = 1$ and the components of its acceleration are, see (1.45),

$$\begin{cases} A^0 = F\,\widetilde{g}_{ab}\,\dfrac{dq^a}{dq^0}\,\dfrac{dq^b}{dq^0}, \\ A^c = \dfrac{d}{dq^0}\,\dfrac{dq^c}{dq^0} + \Gamma^c_{ab}\,\dfrac{dq^a}{dq^0}\,\dfrac{dq^b}{dq^0} + 2\,H\,\dfrac{dq^c}{dq^0}. \end{cases}$$

The symbols A^c are the acceleration components of the projected curve $\widetilde{\gamma}$. Passing to the parameter $\widetilde{s}$ we find:

$$A^0 = F\,\widetilde{g}_{ab}\,\frac{dq^a}{d\widetilde{s}}\,\frac{dq^b}{d\widetilde{s}}\left(\frac{d\widetilde{s}}{dq^0}\right)^2$$

that is, since $\widetilde{g}_{ab}\,\dfrac{dq^a}{d\widetilde{s}}\,\dfrac{dq^b}{d\widetilde{s}} = 1,$

$$A^0 = F\,V^2.$$

Moreover:

$$\begin{aligned} A^c &= \frac{d}{dq^0}\left(V\,\frac{dq^c}{d\widetilde{s}}\right) + \Gamma^c_{ab}\,\frac{dq^a}{d\widetilde{s}}\,\frac{dq^b}{d\widetilde{s}}\,V^2 + 2\,H\,V\,\frac{dq^c}{d\widetilde{s}} \\ &= \frac{dV}{dq^0}\,\frac{dq^c}{d\widetilde{s}} + V\,\frac{d}{dq^0}\left(\frac{dq^c}{d\widetilde{s}}\right) + \Gamma^c_{ab}\,\frac{dq^a}{d\widetilde{s}}\,\frac{dq^b}{d\widetilde{s}}\,V^2 + 2\,H\,V\,\frac{dq^c}{d\widetilde{s}} \\ &= \frac{dV}{dq^0}\,\frac{dq^c}{d\widetilde{s}} + V^2\,\frac{d}{d\widetilde{s}}\left(\frac{dq^c}{d\widetilde{s}}\right) + \Gamma^c_{ab}\,\frac{dq^a}{d\widetilde{s}}\,\frac{dq^b}{d\widetilde{s}}\,V^2 + 2\,H\,V\,\frac{dq^c}{d\widetilde{s}}. \end{aligned}$$

Tidying up:

$$A^c = V^2 \left[\frac{d}{d\tilde{s}} \left(\frac{dq^c}{d\tilde{s}} \right) + \Gamma^c_{ab} \frac{dq^a}{d\tilde{s}} \frac{dq^b}{d\tilde{s}} \right] + \left[\frac{dV}{dq^0} + 2\,H\,V \right] \frac{dq^c}{d\tilde{s}}.$$

First path: requirement 4 $\Longrightarrow$ (i) **and** (ii). For Requirement 4 if γ is a geodesic, i.e. if the equations hold

$$A^0 = \lambda, \quad A^c = \lambda \frac{dq^c}{dq^0} = \lambda \frac{dq^c}{d\tilde{s}}\, V,$$

which as seen above become

$$\begin{cases} F\,V^2 = \lambda, \\ V^2 \left[\dfrac{d}{d\tilde{s}} \left(\dfrac{dq^c}{d\tilde{s}} \right) + \Gamma^c_{ab} \dfrac{dq^a}{d\tilde{s}} \dfrac{dq^b}{d\tilde{s}} \right] + \left[\dfrac{dV}{dq^0} + 2\,H\,V \right] \dfrac{dq^c}{d\tilde{s}} = \lambda \dfrac{dq^c}{d\tilde{s}} V, \end{cases} \tag{1.55}$$

then $\tilde{\gamma}$ must also be a geodesic of the reference metric $\tilde{g}$ and $q^c(\tilde{s})$ must satisfy the equations

$$\frac{d}{d\tilde{s}} \left(\frac{dq^c}{d\tilde{s}} \right) + \tilde{\Gamma}^c_{ab} \frac{dq^a}{d\tilde{s}} \frac{dq^b}{d\tilde{s}} = 0 \tag{1.56}$$

because $\tilde{s}$ is an affine parameter. The same $q^c(\tilde{s})$ must also satisfy the second Eq. (1.55) which, taking into account the first (1.55), become

$$\frac{d}{d\tilde{s}} \left(\frac{dq^c}{d\tilde{s}} \right) + \Gamma^c_{ab} \frac{dq^a}{d\tilde{s}} \frac{dq^b}{d\tilde{s}} = \frac{1}{V} \left[F\,V^2 - \frac{d \log V}{dq^0} - 2\,H \right] \frac{dq^c}{d\tilde{s}}.$$

By subtracting Eq. (1.56) from these equations we find the equality

$$(\Gamma^c_{ab} - \tilde{\Gamma}^c_{ab}) \frac{dq^a}{d\tilde{s}} \frac{dq^b}{d\tilde{s}} = \frac{1}{V} \left[F\,V^2 - \frac{d \log V}{dq^0} - 2\,H \right] \frac{dq^c}{d\tilde{s}}$$

whose first member is homogeneous quadratic in $dq^a/d\tilde{s}$ while the second is linear. Given the arbitrariness of the $dq^a/d\tilde{s}$ the two members must vanish and therefore the following equalities must hold:

$$\Gamma^c_{ab} = \tilde{\Gamma}^c_{ab}, \quad F\,V^2 - \frac{d \log V}{dq^0} - 2\,H = 0.$$

They express the conditions (i) and (ii) of the statement.

Inverse path: (i) **and** (ii) $\Longrightarrow$ **requirement 4**. The equations of the Γ-geodesics (1.45) are

$$\begin{cases} A^0 = \lambda, \\ A^c = \lambda \dfrac{dq^c}{dq^0}, \end{cases} \iff \begin{cases} F\,\widetilde{g}_{ab}\, \dfrac{dq^a}{dq^0}\dfrac{dq^b}{dq^0} = \lambda, \\ \dfrac{d}{dq^0}\dfrac{dq^c}{dq^0} + \Gamma^c_{ab}\, \dfrac{dq^a}{dq^0}\dfrac{dq^b}{dq^0} + 2\,H\,\dfrac{dq^c}{dq^0} = \lambda\,\dfrac{dq^c}{dq^0}. \end{cases}$$

Passing to the parameter $\widetilde{s}$ these equations take on the form (the calculations carried out in the previous path are used)

$$\begin{cases} F\,V^2 = \lambda, \\ \dfrac{d}{d\widetilde{s}}\left(\dfrac{dq^c}{d\widetilde{s}}\right) + \Gamma^c_{ab}\,\dfrac{dq^a}{d\widetilde{s}}\,\dfrac{dq^b}{d\widetilde{s}} = \dfrac{1}{V}\left[\lambda - \dfrac{d\log V}{dq^0} - 2\,H\right]\dfrac{dq^c}{d\widetilde{s}}. \end{cases} \tag{1.57}$$

By hypothesis $\Gamma^c_{ab} = \widetilde{\Gamma}^c_{ab}$ and Eq. (1.54) holds, so the system of Eq. (1.57) translates into

$$\begin{cases} F\,V^2 = \lambda, \\ \dfrac{d}{d\widetilde{s}}\left(\dfrac{dq^c}{d\widetilde{s}}\right) + \widetilde{\Gamma}^c_{ab}\,\dfrac{dq^a}{d\widetilde{s}}\,\dfrac{dq^b}{d\widetilde{s}} = \dfrac{1}{V}\left[\lambda - F\,V^2\right]\dfrac{dq^c}{d\widetilde{s}}. \end{cases}$$

These are the equations of the Γ-geodesics. Question: is Requirement 4 satisfied? That is to say: taken together, do these equations imply that the second ones, which govern the projected curve, are equations of the $\widetilde{\Gamma}$-geodesics? The answer is affirmative because if the second ones are the equations of the $\widetilde{\Gamma}$-geodesics then not only the first member vanishes but also the second one, by virtue of the first equation $F\,V^2 = \lambda$. ■

Remark 1.20 In Eq. (1.54) it is assumed $V > 0$. This means that the two parameters q^0 and $\widetilde{s}$ are considered equi-oriented. •

Remark 1.21 Theorem 1.14 also shows that the equality $\Gamma^c_{ab} = \widetilde{\Gamma}^c_{ab}$ does not depend on the choice of reference space. It follows that, once the spatial co-moving coordinates (q^a) are fixed, *the Christoffel symbols* $\widetilde{\Gamma}^c_{ab}$ *have the same expression in every spatial section.* •

Conclusion. The complete table of symbols of a cosmic connection, that is, of a symmetric linear connection satisfying all the above requirements, turns out to be

$$\boxed{\begin{aligned} &\Gamma^0_{a0} = 0, \quad \Gamma^c_{00} = 0, \quad \Gamma^0_{00} = 0 \\ &\Gamma^c_{a0} = H(q^0)\,\delta^c_a, \quad \Gamma^0_{ab} = F(q^0)\,\widetilde{g}_{ab}(\widetilde{q}), \quad \Gamma^c_{ab} = \widetilde{\Gamma}^c_{ab}. \end{aligned}} \tag{1.58}$$

where the function $F(q^0)$ satisfying the Eq. (1.54) remains indeterminate. This indeterminacy will be resolved with the intervention of a bridge postulate (Chap. 2). From (1.45) it follows that the equations of the geodesics in the parameter q^0 are

$$
\boxed{
\begin{aligned}
& F\,\widetilde{g}_{ab}\,\frac{dq^a}{dq^0}\,\frac{dq^b}{dq^0} = \lambda, \\
& \frac{d}{dq^0}\,\frac{dq^c}{dq^0} + \widetilde{\Gamma}^c_{ab}\,\frac{dq^a}{dq^0}\,\frac{dq^b}{dq^0} = (\lambda - 2\,H)\,\frac{dq^c}{dq^0}.
\end{aligned}
}
\tag{1.59}
$$

1.14 Ricci Tensor of a Cosmic Connection

In this and the following sections we will complete the geometric picture regarding the cosmic space-time by preparing some *ingredients* that will be needed in the formulation of the dynamics of the Universe.

Theorem 1.15 *The components of the Ricci tensor of a cosmic connection* (1.58) *are*

$$
\boxed{
\begin{aligned}
& R_{00} = -3\,(H' + H^2) = -3\,a^{-1}a'' \\
& R_{a0} = 0 \\
& R_{ab} = \widetilde{R}_{ab} + (F' + H\,F)\,\widetilde{g}_{ab} = (F' + H\,F + 2\,\widetilde{K})\,\widetilde{g}_{ab}
\end{aligned}
}
\tag{1.60}
$$

where $\widetilde{K}$ *and* $\widetilde{R}_{ab}$ *are the curvature constant and the Ricci tensor of the reference metric* $\widetilde{g}$.

Trace of the proof. Taking into account Eq. (1.53), the various components defined by the (1.4) are calculated with the symbols (1.58). Note that $R_{a0} = 0$ because the Ricci tensor is necessarily isotropic, see (1.40).

1.15 Covariant Derivatives and Conservation Equations

We denote by ∇_α the covariant derivative associated with a cosmic connection. The components of the derivative of any vector field V^α are given by

$$
\boxed{
\nabla_\alpha V^\beta = \partial_\alpha V^\beta + \Gamma^\beta_{\alpha\gamma}\,V^\gamma
\left\{
\begin{aligned}
& \nabla_0 V^0 = \partial_0 V^0, \\
& \nabla_0 V^b = \partial_0 V^b + H\,V^b, \\
& \nabla_a V^0 = \partial_a V^0 + F\,\widetilde{g}_{ac}\,V^c, \\
& \nabla_a V^b = \partial_a V^b + \widetilde{\Gamma}^b_{ac}\,V^c + H\,\delta^b_a\,V^0
\end{aligned}
\right.
}
\tag{1.61}
$$

So the divergence of this vector is

$$
\nabla_\alpha V^\alpha = \partial_\alpha V^\alpha + \widetilde{\Gamma}^a_{ac}\,V^c + 3\,H\,V^0.
\tag{1.62}
$$

If the vector field is isotropic then (1.38) applies, V^{o} is a function of q^{o} only and $V^a = 0$, therefore:

$$\boxed{\begin{aligned} &\nabla_{\!0} V^{\text{o}} = \partial_0 V^{\text{o}} \\ &\nabla_{\!0} V^b = 0 \\ &\nabla_{\!a} V^{\text{o}} = 0 \\ &\nabla_{\!a} V^b = H\, V^{\text{o}}\, \delta^b_a \end{aligned}} \qquad \boxed{\nabla_{\!\alpha} V^\alpha = \partial_0 V^{\text{o}} + 3\, H\, V^{\text{o}}} \tag{1.63}$$

For a contravariant double symmetric tensor, from the definition

$$\nabla_{\!\alpha} T^{\beta\gamma} = \partial_\alpha T^{\beta\gamma} + \Gamma^\beta_{\alpha\delta} T^{\delta\gamma} + \Gamma^\gamma_{\alpha\delta} T^{\beta\delta}$$

we get the following equations:

$$\begin{cases} \nabla_{\!0} T^{\text{oo}} = \partial_0 T^{\text{oo}} \\ \nabla_{\!0} T^{\text{o}c} = \partial_0 T^{\text{o}c} + H\, T^{\text{o}c} \\ \nabla_{\!0} T^{bc} = \partial_0 T^{bc} + 2\, H\, T^{bc}. \end{cases}$$
$$\begin{cases} \nabla_{\!a} T^{\text{oo}} = \partial_a T^{\text{oo}} + 2\, F\, \tilde{g}_{ab} T^{b\text{o}} \\ \nabla_{\!a} T^{b\text{o}} = \partial_a T^{b\text{o}} + \tilde{\Gamma}^b_{ad} T^{d\text{o}} + F\, \tilde{g}_{ad}\, T^{db} + H\, T^{\text{oo}}\, \delta^b_a \\ \nabla_{\!a} T^{bc} = \partial_a T^{bc} + \tilde{\Gamma}^b_{ad} T^{dc} + \tilde{\Gamma}^c_{ad} T^{bd} + H\, (\delta^b_a\, T^{\text{o}c} + \delta^c_a\, T^{b\text{o}}). \end{cases} \tag{1.64}$$

Therefore, the components of the divergence $\nabla_{\!\alpha} T^{\alpha\beta}$ turns out to be

$$\begin{cases} \nabla_{\!\alpha} T^{\alpha\text{o}} = \nabla_{\!0} T^{\text{oo}} + \nabla_{\!a} T^{a\text{o}} = \partial_0 T^{\text{oo}} + \partial_a T^{a\text{o}} + \tilde{\Gamma}^a_{ad} T^{d\text{o}} + F\, \tilde{g}_{ad}\, T^{da} + 3\, H\, T^{\text{oo}}. \\ \nabla_{\!\alpha} T^{\alpha b} = \nabla_{\!0} T^{\text{o}b} + \nabla_{\!a} T^{ab} = \partial_0 T^{\text{o}b} + H\, T^{\text{o}b} \\ \quad + \partial_a T^{ab} + \tilde{\Gamma}^a_{ad} T^{db} + \tilde{\Gamma}^b_{ad} T^{ad} + H\, (\delta^a_a\, T^{\text{o}b} + \delta^b_a\, T^{a\text{o}}) \\ = \partial_0 T^{\text{o}b} + H\, T^{\text{o}b} + \partial_a T^{ab} + \tilde{\Gamma}^a_{ad} T^{db} + \tilde{\Gamma}^b_{ad} T^{ad} + H\, (3\, T^{\text{o}b} + T^{b\text{o}}). \end{cases} \tag{1.65}$$

If the tensor is isotropic then (1.39) applies, so from the previous formulas we get:

$$\boxed{\begin{aligned} &\nabla_{\!0} T^{\text{oo}} = \phi', \quad \nabla_{\!0} T^{\text{o}c} = 0 \\ &\nabla_{\!0} T^{ab} = \left(\psi' + 2\, H\, \psi\right) \tilde{g}^{ab} \end{aligned}} \tag{1.66}$$

$$\boxed{\begin{aligned} &\nabla_{\!a} T^{\text{oo}} = 0, \quad \nabla_{\!a} T^{b\text{o}} = (H\, \phi + F\, \psi)\, \delta^b_a \\ &\nabla_{\!a} T^{bc} = \psi \left(\partial_a \tilde{g}^{bc} + \tilde{\Gamma}^b_{ad}\, \tilde{g}^{dc} + \tilde{\Gamma}^c_{ad}\, \tilde{g}^{bd}\right) = 0 \end{aligned}} \tag{1.67}$$

$$\boxed{\begin{aligned} \nabla_\alpha T^{\alpha 0} &= \phi' + 3\,(H\,\phi + F\,\psi) \\ \nabla_\alpha T^{\alpha b} &= 0 \end{aligned}} \tag{1.68}$$

These last equations prove a property that will have a notable consequence in dealing with relativistic cosmology (Theorem 2.7).

Theorem 1.16 *With respect to any cosmic connection for any symmetric double covariant tensor $T^{\alpha\beta}$ the four conservation equations $\nabla_\alpha T^{\alpha\beta} = 0$ are equivalent to a single equation:*

$$\boxed{\phi' + 3\,(H\,\phi + F\,\psi) = 0} \tag{1.69}$$

Chapter 2
Bridge-Postulates

As mentioned in the Preface, a bridge-postulate allows to transit from the purely geometric territory concerning the structure of cosmic space-time to the territory of a **cosmic dynamics** where the equations governing the evolution of the scale factor $a(t)$ will be determined.

A bridge-postulate has the task of finding, among the infinite possible ones, a single cosmic connection, i.e., a single function $F(q^0)$ that completes the table of symbols (1.58). A connection is the indispensable tool for defining the concept of **acceleration** and also for writing field equations.

In addition, with the assignment of a connection, cosmic time t, which has so far remained an indeterminate parameter, will acquire physical meaning. In other words, it will be possible to define a **standard clock** with which to measure it.

2.1 Newtonian Cosmic Connection

Newtonian bridge-postulate. *Cosmic time t is an affine parameter for free particle histories.*

In a cosmic connection time t is an affine parameter of galactic histories (Requirement 2). This postulate extends this property to free particles.

Theorem 2.1 (i) *There is only one cosmic connection satisfying the Newtonian bridge-postulate and it is characterized by the condition $F = 0$.* (ii) *The equations of the transversal geodesics to the space sections are*

$$\boxed{\frac{d}{dt}\frac{dq^c}{dt} + \widetilde{\Gamma}^c_{ab}\,\frac{dq^a}{dt}\,\frac{dq^b}{dt} = -2\,H(t)\,\frac{dq^c}{dt}} \tag{2.1}$$

S. Benenti, *Mathematical Foundations and Numerical Analysis of the Dynamics of an Isotropic Universe*, https://doi.org/10.1007/978-3-031-64033-9_2

Proof (i) Let us rewrite Eq. (1.59) of the geodesics of a cosmic connection taking cosmic time t as parameter,

$$\begin{cases} F\,\widetilde{g}_{ab}\,\dfrac{dq^a}{dt}\,\dfrac{dq^b}{dt} = \lambda\,\kappa^2, \\ \dfrac{d}{dt}\dfrac{dq^c}{dt} + \widetilde{\Gamma}^c_{ab}\,\dfrac{dq^a}{dt}\,\dfrac{dq^b}{dt} = \kappa\,\left(\lambda - 2\,H(q^0)\right)\,\dfrac{dq^c}{dt}. \end{cases}$$

The postulate is equivalent to the condition $\lambda = 0$ and the first equation shows that this is satisfied if and only if $F = 0$. (ii) The second set of geodesic equations reduces to

$$\frac{d}{dt}\frac{dq^c}{dt} + \widetilde{\Gamma}^c_{ab}\,\frac{dq^a}{dt}\,\frac{dq^b}{dt} = -2\,\kappa\,H(q^0)\,\frac{dq^c}{dt}.$$

By virtue of (1.53) $H(t) = \kappa\,H(q^0)$, we get Eq. (2.1). ■

The space-time geodesics of a cosmic connection are projected into geodesics of the reference space (Requirement 4). Equations (2.1) are the equations of the projections of the space-time geodesics of free particles. At the left hand side we find the a^c components of the acceleration with respect to the cosmic time t of a moving point in reference space. The right hand side can then be interpreted as a *force*

$$f^c = -2\,H(t)\,\frac{dq^c}{dt} \tag{2.2}$$

acting on this point and directed according to its velocity, concordant for $H(t) < 0$, opposite for $H(t) > 0$. This fact is not in agreement with classical Newtonian dynamics where, by principle, a free particle (i.e. not submitted to external stresses) moves in rectilinear and uniform motion. Concordance occurs for $H = 0$ that is, for $a(t) =$ constant. This is the case of a *static Universe*.

We call this connection **Newtonian** since a cosmic spacetime equipped with this connection is a generalization of the Newtonian space-time of classical mechanics, where:

1. The manifold M is an affine four-dimensional space.
2. The spatial sections are Euclidean three-dimensional affine spaces.
3. The cosmic world-lines are parallel straight lines and represent the motion of the so-called **fixed stars**. The congruence of these lines is an **inertial reference frame**, as well as any other congruence of parallel lines transversal to the foliation S_t.
4. The world-lines of the free-falling particles are transversal straight lines (**law of inertia**).
5. The cosmic time t is the **absolute time**.
6. The expansion factor $a(t)$ is constant and equal to 1, and the Hubble parameter vanishes: $H = 0$. Consequently, if the space-like coordinates are Cartesian, then

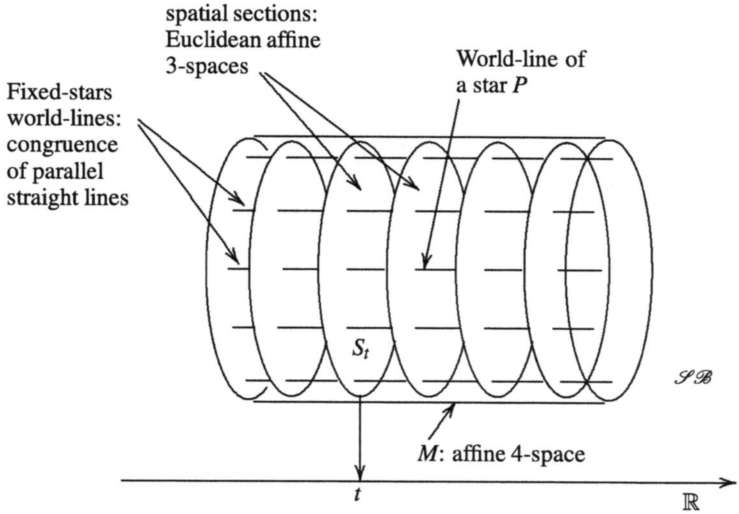

Fig. 2.1 Newtonian space-time

all symbols $\Gamma^{\gamma}_{\alpha\beta}$ vanish. The cosmic connection is flat and coincides with the canonical connection of an affine space (Fig. 2.1).

A Newtonian connection opens the way to the development of Newtonian cosmic dynamical models.

2.2 Peculiar Velocity of a Wandering Particle

In the next section, a second bridge-postulate will be proposed. To do this we need some premises. Let $(q^0(t), q^a(t))$ be the parametric equations of the history $\gamma(t)$ of a wandering particle. When at time t it is passing through a galaxy B_t its bf cosmic velocity is the vector $\boldsymbol{V}$ with components

$$\boldsymbol{V}(t) = \left[\frac{dq^0}{dt}, \frac{dq^a}{dt}\right] \tag{2.3}$$

where the vector

$$\boldsymbol{v}(t) = \left[\frac{dq^a}{dt}\right]$$

is tangent to the spatial section S_t. This is the velocity vector of the particle nearby the galaxy B_t interpreted as a physical reference space and defined by the three vectors associated with the coordinates (q^a), see Fig. 1.7.

We call the length of this spatial vector the **peculiar velocity** of the wandering particle:

$$\boxed{v_{\mathrm{pec}}(t) \stackrel{\mathrm{def}}{=} \sqrt{g_{ab}(t)\,\frac{dq^a}{dt}\,\frac{dq^b}{dt}}} \tag{2.4}$$

Because of what was said above, this is the **relative scalar velocity** in the reference B_t. By virtue of the factorization Eq. (1.26) it takes the form

$$\boxed{v_{\mathrm{pec}}(t) = a(t, t_\sharp)\sqrt{g_{\sharp ab}\,\frac{dq^a}{dt}\,\frac{dq^b}{dt}}} \tag{2.5}$$

$t_{t_\sharp}$ being the reference time of the scale factor and $g_{\sharp ab}$ the components of the metric tensor of the reference space $S_{t_\sharp}$.

If we consider the arc-element $d\widetilde{s}$ of the reference metric $g_\sharp$, defined by the equation

$$g_{\sharp ab}\,\frac{dq^a}{d\widetilde{s}}\,\frac{dq^b}{d\widetilde{s}} = 1,$$

then the peculiar velocity (2.5) takes the simple form

$$\boxed{v_{\mathrm{pec}}(t) = a(t, t_\sharp)\,\frac{d\widetilde{s}}{dt}} \tag{2.6}$$

2.3 Relativistic Cosmic Connection

Relativistic bridge-postulate. *There exists a cosmic time t and there exist free particles whose peculiar velocity with respect to this time t is a universal constant c*: $v_{\mathrm{pec}}(t) = c$.

By virtue of (2.6) equation $v_{\mathrm{pec}}(t) = c$ is equivalent to

$$\boxed{a(t, t_\sharp)\,\frac{d\widetilde{s}}{dt} = c} \tag{2.7}$$

Although with abuse of language, we will call **photons** these particles.

Theorem 2.2 *The existence of photons is compatible with a cosmic connection if and only if*

$$\boxed{F(q^0) = \frac{\kappa^2}{c^2}\,a\,a'} \tag{2.8}$$

Proof By the relativistic postulate a photon is a free particle and by definition of a free particle its history is a geodesic of the cosmic connection. By Requirement 4 this is projected onto a geodesic of the reference space and furthermore, by Theorem 1.14, along this geodesic Eq. (1.54) holds:

$$[*] \quad \frac{d\log V}{dq^{0}} + 2\,H(q^{0}) = F(q^{0})\,V^{2}, \quad V(q^{0}) \overset{\text{def}}{=} \frac{d\widetilde{s}}{dq^{0}}.$$

By virtue of the expression (2.6) of the peculiar velocity and of equality $v_{rmpec}(t) = c$ following from the postulate we have

$$V = \frac{d\widetilde{s}}{dq^{0}} = \frac{1}{\kappa}\frac{d\widetilde{s}}{dt} = \frac{1}{\kappa\,a}\,v_{\text{pec}} = \frac{c}{\kappa\,a}.$$

Hence, c and κ being constants,

$$\frac{d\log V}{dq^{0}} = -\frac{d\log a}{dq^{0}} = -\frac{a'}{a},$$

so that [∗] becomes

$$-\frac{a'}{a} + 2\,H(q^{0}) = F\,\frac{c^2}{\kappa^2\,a^2}.$$

Since $H = a'/a$, by the definition (1.52), the result is

$$\frac{a'}{a} = F\,\frac{c^2}{\kappa^2\,a^2}.$$

From here follows (2.8). ∎

Recalling the table (1.58), it follows from this theorem that

Theorem 2.3 *The relativistic bridge-postulate determines a unique* **relativistic cosmic connection** *whose symbols are*

$$\boxed{\begin{array}{c} \Gamma^{0}_{00} = 0, \quad \Gamma^{0}_{a0} = 0, \quad \Gamma^{c}_{00} = 0 \\ \Gamma^{c}_{a0} = \dfrac{a'}{a}\,\delta^{c}_{a}, \quad \Gamma^{0}_{ab} = \dfrac{\kappa^2}{c^2}\,a\,a'\,\widetilde{g}_{ab}, \quad \Gamma^{c}_{ab} = \widetilde{\Gamma}^{c}_{ab}(\widetilde{q}) \end{array}} \tag{2.9}$$

Along any history parameterized by cosmic time t these symbols should be regarded as functions of t. Since

$$a' = \dot{a}\,\frac{dt}{dq^{0}} = \frac{\dot{a}}{\kappa},$$

we have

$$\boxed{\begin{array}{c} \Gamma^{0}_{00} = 0, \quad \Gamma^{0}_{a0} = 0, \quad \Gamma^{c}_{00} = 0 \\ \Gamma^{c}_{a0} = \frac{1}{\kappa}\frac{\dot{a}}{a}\,\delta^{c}_{a}, \quad \Gamma^{0}_{ab} = \frac{\kappa}{c^2}\,a\,\dot{a}\,\widetilde{g}_{ab}, \quad \Gamma^{c}_{ab} = \widetilde{\Gamma}^{c}_{ab}(\widetilde{q}) \end{array}} \tag{2.10}$$

With these symbols, the equations of geodesics with parameter t

$$\frac{d^2q^\gamma}{dt^2} + \Gamma^{\gamma}_{\alpha\beta}\,\frac{dq^\alpha}{dt}\,\frac{dq^\beta}{dt} = \lambda\,\frac{dq^\gamma}{dt}$$

break into the system

$$\begin{cases} \gamma = 0 \implies \dfrac{d^2q^0}{dt^2} + \Gamma^{0}_{\alpha\beta}\,\dfrac{dq^\alpha}{dt}\,\dfrac{dq^\beta}{dt} = \lambda\,\dfrac{dq^0}{dt}, \\ \gamma = c \implies \dfrac{d^2q^c}{dt^2} + \Gamma^{c}_{\alpha\beta}\,\dfrac{dq^\alpha}{dt}\,\dfrac{dq^\beta}{dt} = \lambda\,\dfrac{dq^c}{dt}. \end{cases}$$

$$\implies \begin{cases} \dfrac{d^2q^0}{dt^2} + \dfrac{\kappa}{c^2}\,a\,\dot{a}\,\widetilde{g}_{ab}\,\dfrac{dq^\alpha}{dt}\,\dfrac{dq^\beta}{dt} = \lambda\,\dfrac{dq^0}{dt}, \\ \dfrac{d^2q^c}{dt^2} + \widetilde{\Gamma}^{c}_{ab}\,\dfrac{dq^a}{dt}\,\dfrac{dq^b}{dt} + \dfrac{2}{\kappa}\dfrac{\dot{a}}{a}\,\dfrac{dq^0}{dt}\,\dfrac{dq^c}{dt} = \lambda\,\dfrac{dq^c}{dt}. \end{cases}$$

Since $\dfrac{dq^0}{dt} = \kappa \implies \begin{cases} \dfrac{\kappa}{c^2}\,a\,\dot{a}\,\widetilde{g}_{ab}\,\dfrac{dq^\alpha}{dt}\,\dfrac{dq^\beta}{dt} = \lambda\,\kappa, \\ \dfrac{d^2q^c}{dt^2} + \widetilde{\Gamma}^{c}_{ab}\,\dfrac{dq^a}{dt}\,\dfrac{dq^b}{dt} + 2\,\dfrac{\dot{a}}{a}\,\dfrac{dq^c}{dt} = \lambda\,\dfrac{dq^c}{dt}. \end{cases}$

$$\implies \begin{cases} \dfrac{a\,\dot{a}}{c^2}\,\widetilde{g}_{ab}\,\dfrac{dq^a}{dt}\,\dfrac{dq^b}{dt} = \lambda, \\ \dfrac{d^2q^c}{dt^2} + \widetilde{\Gamma}^{c}_{ab}\,\dfrac{dq^a}{dt}\,\dfrac{dq^b}{dt} = \left(\lambda - 2\,\dfrac{\dot{a}}{a}\right)\dfrac{dq^c}{dt}. \end{cases} \tag{2.11}$$

The first equation shows that it cannot be $\lambda = 0$. This means that.

Theorem 2.4 *Cosmic time is not an affine parameter for transverse geodesics, so neither is it for photon histories.*

Further properties.

Theorem 2.5 *The geodesic equations of the relativistic cosmic connection* (2.11) *reduce to the three equations*

$$\boxed{\frac{d^2q^c}{dt^2} + \widetilde{\Gamma}^{c}_{ab}\,\frac{dq^a}{dt}\,\frac{dq^b}{dt} = -H(t)\,\frac{dq^c}{dt}} \tag{2.12}$$

Proof Since $v_{\text{pec}} = a \dfrac{d\widetilde{s}}{dt}$, from the first Eq. (2.11) we derive the multiplier:

$$\lambda = \frac{a\,\dot{a}}{c^2}\,\widetilde{g}_{ab}\,\frac{dq^a}{d\widetilde{s}}\,\frac{dq^b}{d\widetilde{s}}\left(\frac{d\widetilde{s}}{dt}\right)^2 = \frac{a\,\dot{a}}{c^2}\left(\frac{d\widetilde{s}}{dt}\right)^2 = \frac{\dot{a}}{a\,c^2}\,v_{\text{pec}}^2 = \frac{\dot{a}}{a}.$$

Substituting this expression into the second Eq. (2.11) we find (2.12). ■

Remark 2.1 The auxiliary velocity κ does not appear in these equations. ●

Theorem 2.6 *If $t_\omega = +\infty$ and if the scale factor $a(t)$ is unbounded for $t \to +\infty$ then the photon histories asymptotically approach the cosmic fluid histories.*

Proof Since Eq. (2.7) holds for any t, the limit $\lim_{t\to+\infty} a(t) = +\infty$ implies

$$\lim_{t\to+\infty} \frac{d\widetilde{s}}{dt} = 0.$$

This means that in the cosmic monitor, the trace of a photon tends asymptotically to a fixed point. ■

Remark 2.2 This theorem shows a kinematic property of photons independent of the choice of dynamical postulates. ●

Additional properties of the relativistic cosmic connection concern covariant derivatives. From the general formulas (1.66) and (1.69) it follows that

Theorem 2.7 (i) *With respect to the relativistic cosmic connection* (2.9) *the covariant derivatives of an isotropic vector field are*

$$\boxed{\begin{aligned} &\nabla_0 V^0 = \partial_0 V^0, \quad \nabla_0 V^a = 0, \quad \nabla_b V^0 = 0, \\ &\nabla_b V^a = \frac{a'}{a}\, V^0\, \delta^a_b. \end{aligned}} \tag{2.13}$$

(ii) *The divergence is given by*

$$\boxed{\nabla_\alpha V^\alpha = \partial_0 V^0 + 3\,\frac{a'}{a}\,V^0} \tag{2.14}$$

(iii) *For each symmetric contravariant two-tensor $T^{\alpha\beta}$ the four conservation equations $\nabla_\alpha T^{\alpha\beta} = 0$ are equivalent to the single equation*

$$\boxed{\phi' + 3\,a'\left(\frac{\phi}{a} + \frac{\kappa^2}{c^2}\,a\,\psi\right) = 0} \tag{2.15}$$

2.4 Relativistic Cosmic Metric

Here we discover that the relativistic connection is the Levi-Civita connection of a Lorentzian metric in cosmic space-time.

Theorem 2.8 *The relativistic cosmic connection is the Levi-Civita connection of the Lorentzian metric*

$$\boxed{g_{\alpha\beta}\, dq^\alpha\, dq^\beta = \alpha \left(dq^{0\,2} - \frac{\kappa^2 a^2}{c^2}\, \widetilde{g}_{ab}\, dq^a\, dq^b \right)} \tag{2.16}$$

$$\begin{cases} g_{00} = \alpha \\ g_{0a} = 0 \\ g_{ab} = -\alpha\, \dfrac{\kappa^2}{c^2}\, a^2\, \widetilde{g}_{ab} \end{cases}$$

where α is a non-zero constant factor and $q^0 = \kappa\, t$.

Remark 2.3 Observe that, by virtue of the factorization (1.26), the spatial components g_{ab} do not depend on the choice of the reference space, i.e., on the reference time of the scale factor. ●

Remark 2.4 By fixing $\alpha = -1$ we obtain the Lorentzian metric

$$\boxed{g_{\alpha\beta}\, dq^\alpha\, dq^\beta = -\, dq^{0\,2} + \frac{\kappa^2}{c^2}\, a^2\, \widetilde{g}_{ab}\, dq^a\, dq^b} \tag{2.17}$$

i.e.

$$\begin{cases} g_{00} = -1, \\ g_{0a} = 0, \\ g_{ab} = \dfrac{\kappa^2}{c^2}\, a^2\, \widetilde{g}_{ab}. \end{cases}$$

Proof According to Theorem 1.8 the components of any isotropic metric tensor must be of the type

$$g_{\alpha\beta} : \begin{cases} g_{00} = \alpha(q^0), \\ g_{0a} = 0, \\ g_{ab} = \beta(q^0)\, \widetilde{g}_{ab}(\widetilde{q}), \end{cases} \tag{2.18}$$

i.e.

$$g_{\alpha\beta}\, dq^\alpha\, dq^\beta = \alpha\, dq^{0\,2} + \beta\, \widetilde{g}_{ab}\, dq^a\, dq^b.$$

(i) Computation of the first kind Christoffel symbols.

$$\Gamma_{\alpha\beta,\gamma} = \tfrac{1}{2}\left(\partial_\alpha g_{\beta\gamma} + \partial_\beta g_{\gamma\alpha} - \partial_\gamma g_{\alpha\beta}\right).$$

$$2\,\Gamma_{00,\gamma} = \partial_0 g_{0\gamma} + \partial_0 g_{\gamma 0} - \partial_\gamma g_{00} = \begin{cases} 2\,\Gamma_{00,0} = \partial_0 g_{00} + \partial_0 g_{00} - \partial_0 g_{00} = \alpha'. \\ 2\,\Gamma_{00,a} = \partial_0 g_{0a} + \partial_0 g_{a0} - \partial_a g_{00} = 0. \end{cases}$$

$$2\,\Gamma_{0b,\gamma} = \partial_0 g_{b\gamma} + \partial_b g_{\gamma 0} - \partial_\gamma g_{0b} = \begin{cases} 2\,\Gamma_{0b,0} = \partial_0 g_{b0} + \partial_b g_{00} - \partial_0 g_{0b} = 0. \\ 2\,\Gamma_{0b,c} = \partial_0 g_{bc} + \partial_b g_{c0} - \partial_c g_{0b} = \beta'\,\widetilde{g}_{bc}. \end{cases}$$

$$2\,\Gamma_{ab,\gamma} = \partial_a g_{b\gamma} + \partial_b g_{\gamma a} - \partial_\gamma g_{ab} = \begin{cases} 2\,\Gamma_{ab,0} = \partial_a g_{b0} + \partial_b g_{0a} - \partial_0 g_{ab} = -\beta'\,\widetilde{g}_{ab}. \\ 2\,\Gamma_{ab,c} = \partial_a g_{bc} + \partial_b g_{ca} - \partial_c g_{ab} = 2\,\beta\widetilde{\Gamma}_{ab,c}. \end{cases}$$

Non-identically null symbols:

$$\begin{cases} \Gamma_{00,0} = \tfrac{1}{2}\,\alpha', \\ \Gamma_{0b,c} = \tfrac{1}{2}\,\beta'\,\widetilde{g}_{bc}, \end{cases} \qquad \begin{cases} \Gamma_{ab,0} = -\tfrac{1}{2}\,\beta'\,\widetilde{g}_{ab}, \\ \Gamma_{ab,c} = \beta\,\widetilde{\Gamma}_{ab,c}. \end{cases}$$

(ii) Computation of the second kind Christoffel symbols. $\Gamma^\gamma_{\alpha\beta} = g^{\gamma\delta}\,\Gamma_{\alpha\beta,\delta}$.

$$g^{\alpha\beta} : \begin{cases} g^{00} = \alpha^{-1}(q^0), \\ g^{0a} = 0, \\ g^{ab} = \beta^{-1}(q^0)\,\widetilde{g}^{ab}(\widetilde{q}). \end{cases}$$

$$\Gamma^\gamma_{00} = g^{\gamma\delta}\,\Gamma_{00,\delta} = \begin{cases} \Gamma^0_{00} = g^{0\delta}\,\Gamma_{00,\delta} = g^{00}\,\Gamma_{00,0} = \tfrac{1}{2}\,\alpha^{-1}\,\alpha' = \tfrac{1}{2}\,(\log\alpha)'. \\ \Gamma^c_{00} = g^{c\delta}\,\Gamma_{00,\delta} = g^{cd}\,\Gamma_{00,d} = 0. \end{cases}$$

$$\Gamma^\gamma_{a0} = g^{\gamma\delta}\,\Gamma_{a0,\delta} = \begin{cases} \Gamma^0_{a0} = g^{0\delta}\,\Gamma_{a0,\delta} = g^{00}\,\Gamma_{a0,0} = 0. \\ \Gamma^c_{a0} = g^{c\delta}\,\Gamma_{a0,\delta} = g^{cd}\,\Gamma_{a0,d} = \tfrac{1}{2}\,\beta^{-1}\,\widetilde{g}^{cd}\,\beta'\,\widetilde{g}_{ad} \\ \quad = \tfrac{1}{2}\,(\log\beta)'\,\delta^c_a. \end{cases}$$

$$\Gamma^\gamma_{ab} = g^{\gamma\delta}\,\Gamma_{ab,\delta} = \begin{cases} \Gamma^0_{ab} = g^{0\delta}\,\Gamma_{ab,\delta} = g^{00}\,\Gamma_{ab,0} = -\tfrac{1}{2}\,\alpha^{-1}\,\beta'\,\widetilde{g}_{ab}. \\ = g^{c\delta}\,\Gamma_{ab,\delta} = g^{cd}\,\Gamma_{ab,d} = \beta^{-1}\,\widetilde{g}^{cd}\,\beta\,\widetilde{\Gamma}_{ab,d} = \widetilde{\Gamma}^c_{ab}. \end{cases}$$

Summary overview:

$$\boxed{\begin{array}{c} \Gamma^0_{00} = \tfrac{1}{2}\,\dfrac{\alpha'}{\alpha}, \quad \Gamma^0_{a0} = 0, \quad \Gamma^c_{00} = 0 \\ \Gamma^c_{a0} = \tfrac{1}{2}\,\dfrac{\beta'}{\beta}\,\delta^c_a, \quad \Gamma^0_{ab} = -\tfrac{1}{2}\,\dfrac{\beta'}{\alpha}\,\widetilde{g}_{ab}, \quad \Gamma^c_{ab} = \widetilde{\Gamma}^c_{ab}(\widetilde{q}) \end{array}}$$

These symbols coincide with those of the relativistic cosmic connection (2.9)

$$\begin{cases} \Gamma^{0}_{00} = 0, \quad \Gamma^{0}_{a0} = 0, \quad \Gamma^{c}_{00} = 0, \\ \Gamma^{c}_{a0} = \dfrac{a'}{a}\,\delta^{c}_{a}, \quad \Gamma^{0}_{ab} = \dfrac{\kappa^2}{c^2}\,a\,a'\,\widetilde{g}_{ab}, \quad \Gamma^{c}_{ab} = \widetilde{\Gamma}^{c}_{ab}(\widetilde{q}), \end{cases}$$

if and only if

$$[*] \quad \begin{cases} \alpha = \text{constant} \\ \frac{1}{2}\,(\log\beta)' = (\log a)' \\ -\frac{1}{2}\,\alpha^{-1}\,\beta' = \dfrac{\kappa^2}{c^2}\,a\,a' \end{cases}$$

Hence,

$$[*] \Longrightarrow \begin{cases} \alpha = \text{constant} \\ a^{-2}\,\beta = \text{constant} = \gamma \\ \beta' = -2\,\alpha\,\dfrac{\kappa^2}{c^2}\,a\,a' \end{cases} \Longrightarrow \begin{cases} \alpha = \text{constant} \\ \beta = \gamma\,a^2 \\ \beta' = -2\,\alpha\,\dfrac{\kappa^2}{c^2}\,a\,a' \end{cases}$$

$$\Longrightarrow \begin{cases} \alpha = \text{constant} \\ \beta' = 2\,\gamma\,a\,a' \\ \beta' = -2\,\alpha\,\dfrac{\kappa^2}{c^2}\,a\,a' \end{cases} \Longrightarrow \begin{cases} \alpha = \text{constant} \\ 2\,\gamma\,a\,a' = -2\,\alpha\,\dfrac{\kappa^2}{c^2}\,a\,a' \end{cases} \Longrightarrow \begin{cases} \alpha = \text{constant} \\ \gamma = -\,\alpha\,\dfrac{\kappa^2}{c^2} \\ \beta = \gamma\,a^2 = -\,\alpha\,\dfrac{\kappa^2}{c^2}\,a^2. \end{cases}$$

Comparison with (2.18) proves (2.16). ■

Theorem 2.9 *In the metric* (2.17) (i) *the cosmic fluid histories are time-like geodesics orthogonal to the spatial sections* S_t *and* (ii) *the photon histories are null (light-like) geodesics.*

Proof (i) For the first requirement of a cosmic connection the cosmic fluid histories are geodesics of the relativistic connection, hence of the cosmic metric. Since in co-moving coordinates $g_{0a} = 0$, these histories are orthogonal to the spatial sections so they are time-like. (ii) For each history parameterized by q^0 we have

$$\begin{aligned} g_{\alpha\beta}\,\frac{dq^\alpha}{dq^0}\,\frac{dq^\alpha}{dq^0} &= -1 + \frac{\kappa^2}{c^2}\,a^2\,\widetilde{g}_{ab}\,\frac{dq^a}{dq^0}\,\frac{dq^b}{dq^0} \\ &= -1 + \frac{\kappa^2}{c^2}\,a^2\,\widetilde{g}_{ab}\,\frac{dq^a}{d\widetilde{s}}\,\frac{dq^b}{d\widetilde{s}}\left(\frac{d\widetilde{s}}{dq^0}\right)^2 = -1 + \frac{\kappa^2}{c^2}\,a^2\left(\frac{d\widetilde{s}}{dq^0}\right)^2 \end{aligned} \tag{2.19}$$

being $\widetilde{g}_{ab}\,\dfrac{dq^a}{d\widetilde{s}}\,\dfrac{dq^b}{d\widetilde{s}} = 1$. On the other hand, we have seen that the equation $v_{\text{pec}}(t) = c$, which by postulate assigns to a photon a constant peculiar velocity c, is equivalent to Eq. (2.7), which in turns can be put in the form

$$\frac{d\widetilde{s}}{dt} = \frac{c}{a(t, t_\sharp)}.$$

Since we have defined $q^0 = \kappa t$, it follows that $\dfrac{d\widetilde{s}}{dq^0} = \dfrac{d\widetilde{s}}{\kappa\, dt} = \dfrac{c}{\kappa\, a(t, t_\sharp)}$, and we see that the last term in (2.19) vanishes. This proves that the curve is light-like. It is a geodesic because, by postulate, a photon is a free particle. ■

2.5 Ricci and Einstein Tensors

The components of the Ricci tensor of the relativistic cosmic connection are

$$\boxed{\begin{aligned} R_{00} &= -3\, a^{-1} a'', \quad R_{a0} = 0 \\ R_{ab} &= \widetilde{R}_{ab} + \frac{\kappa^2}{c^2} \left(2\, a'^2 + a\, a''\right) \widetilde{g}_{ab} \\ &= \left(\frac{\kappa^2}{c^2} \,(2\, a'^2 + a\, a'') + 2\, \widetilde{K}\right) \widetilde{g}_{ab} \end{aligned}} \tag{2.20}$$

These expressions follow from the (1.60) by posing

$$H(q^0) = \frac{a'}{a}, \quad F = \frac{\kappa^2}{c^2}\, a\, a', \quad F' = \frac{\kappa^2}{c^2}\, (a'^2 + a\, a'').$$

We can calculate the mixed and contravariant components of the Ricci tensor by raising the indices of the covariant components (2.20) by means of the contravariant components of the metric tensor (2.17):

$$g^{\alpha\beta}: \begin{cases} g^{00} = -1, \\ g^{0a} = 0, \\ g^{ab} = \dfrac{c^2}{\kappa^2} \dfrac{1}{a^2}\, \widetilde{g}^{ab}. \end{cases} \tag{2.21}$$

The resulting expressions are

$$\boxed{\begin{aligned} R^0_0 &= 3\, \frac{a''}{a}, \quad R^0_a = 0, \quad R^b_0 = 0, \\ R^b_a &= \frac{1}{a^2} \left[\frac{c^2}{\kappa^2}\, \widetilde{R}^b_a + \left(2\, a'^2 + a\, a''\right)\, \delta^b_a\right] \\ &= \frac{1}{a^2} \left[2\, a'^2 + a\, a'' + 2\, \frac{c^2}{\kappa^2}\, \widetilde{K}\right] \delta^b_a \end{aligned}} \tag{2.22}$$

$$\boxed{\begin{aligned} R^{00} &= -3\,\frac{a''}{a}, \quad R^{a0} = 0, \\ R^{ab} &= \frac{c^2}{\kappa^2}\,\frac{1}{a^4}\left[\frac{c^2}{\kappa^2}\,\widetilde{R}^{ab} + \left(2\,a'^2 + a\,a''\right)\,\widetilde{g}^{ab}\right] \\ &= \frac{c^2}{\kappa^2}\,\frac{1}{a^4}\left[2\,a'^2 + a\,a'' + 2\,\frac{c^2}{\kappa^2}\,\widetilde{K}\right]\widetilde{g}^{ab}. \end{aligned}} \tag{2.23}$$

$$\boxed{\begin{aligned} R &\overset{\text{def}}{=} R^{\alpha}_{\alpha} = \frac{3}{a^2}\left[\frac{c^2}{\kappa^2}\,\widetilde{R} + 2\,\left(a'^2 + a\,a''\right)\right] \\ &= \frac{6}{a^2}\left(a'^2 + a\,a'' + \frac{c^2}{\kappa^2}\,\widetilde{K}\right) \end{aligned}} \tag{2.24}$$

Along any curve parameterized by cosmic time t all tensor components must be expressed as functions of t. Since

$$a' = \frac{\dot{a}}{\kappa},\ a'' = \frac{\ddot{a}}{\kappa^2},$$

we get:

$$\boxed{\begin{aligned} R^{0}_{0} &= \frac{3}{\kappa^2}\,\frac{\ddot{a}}{a}, \quad R^{0}_{a} = 0, \quad R^{b}_{0} = 0, \\ R^{b}_{a} &= \frac{1}{\kappa^2\,a^2}\left[c^2\,\widetilde{R}^{b}_{a} + \left(2\,\dot{a}^2 + a\,\ddot{a}\right)\,\delta^{b}_{a}\right] \\ &= \frac{1}{\kappa^2\,a^2}\left[2\,\dot{a}^2 + a\,\ddot{a} + 2\,c^2\,\widetilde{K}\right]\delta^{b}_{a} \end{aligned}} \tag{2.25}$$

$$\boxed{\begin{aligned} R^{00} &= -\frac{3}{\kappa^2}\,\frac{\ddot{a}}{a}, \quad R^{a0} = 0, \\ R^{ab} &= \frac{c^2}{\kappa^4}\,\frac{1}{a^4}\left[c^2\,\widetilde{R}^{ab} + \left(2\,\dot{a}^2 + a\,\ddot{a}\right)\,\widetilde{g}^{ab}\right] \\ &= \frac{c^2}{\kappa^4}\,\frac{1}{a^4}\left[2\,\dot{a}^2 + a\,\ddot{a} + 2\,c^2\,\widetilde{K}\right]\widetilde{g}^{ab}. \end{aligned}} \tag{2.26}$$

$$\boxed{R = \frac{3}{\kappa^2\,a^2}\left[c^2\,\widetilde{R} + 2\,\left(\dot{a}^2 + a\,\ddot{a}\right)\right] = \frac{6}{\kappa^2\,a^2}\left(\dot{a}^2 + a\,\ddot{a} + c^2\,\widetilde{K}\right)} \tag{2.27}$$

It follows that the contravariant components of the Einstein tensor

$$G^{\alpha\beta} \overset{\text{def}}{=} R^{\alpha\beta} - \tfrac{1}{2}\,R\,g^{\alpha\beta}$$

are

$$\boxed{\begin{aligned} G^{00} &= \frac{3}{a^2}\left(a'^2 + \frac{c^2}{\kappa^2}\widetilde{K}\right), \quad G^{a0} = 0, \\ G^{ab} &= -\frac{c^2}{\kappa^2 a^4}\left(a'^2 + 2\,a\,a'' + \frac{c^2}{\kappa^2}\widetilde{K}\right)\widetilde{g}^{ab} \end{aligned}} \tag{2.28}$$

$$\boxed{\begin{aligned} G^{00} &= \frac{3}{\kappa^2 a^2}\left(\dot{a}^2 + c^2\,\widetilde{K}\right), \quad G^{a0} = 0, \\ G^{ab} &= -\frac{c^2}{\kappa^4 a^4}\left(\dot{a}^2 + 2\,a\,\ddot{a} + c^2\,\widetilde{K}\right)\widetilde{g}^{ab} \end{aligned}} \tag{2.29}$$

The auxiliary constant κ, which has the dimension of a velocity, was introduced to define a length-dimensional coordinate $q^0 = \kappa\, t$ as a substitute of cosmic time. Its numerical value is indeterminate. So far this constant has proved to be a useful tool, especially in checking the dimensional homogeneity of many equations. On the threshold of relativistic dynamics it is quite natural to place this constant equal to the speed of light c:

$$\boxed{\kappa = c} \qquad \boxed{q^0 = c\,t}$$

The cosmic metric then takes the expression

$$\boxed{g_{\alpha\beta}\,dq^\alpha\,dq^\beta = -\,c^2\,dt^2 + a^2\,\widetilde{g}_{ab}\,dq^a\,dq^b} \tag{2.30}$$

2.6 Sub-luminal Particles

We call **sub-luminal particle** a wandering particle whose history $q^\alpha(t)$ is time-like in the metric (2.30). Every tiime-like curve admits a parameter τ, called **proper time**, such that

$$\boxed{g_{\alpha\beta}\,\frac{dq^\alpha}{d\tau}\,\frac{dq^\beta}{d\tau} = -\,c^2} \tag{2.31}$$

The four-dimensional vector

$$\boxed{V(\tau) \overset{\text{def}}{=} \left[\frac{dq^\alpha}{d\tau}\right]} \tag{2.32}$$

is called the **proper velocity** or **absolute velocity** of the particle.

Theorem 2.10 *Sub-luminal particles have peculiar velocity lower than the universal constant c.*

Proof Let us recall the definition of peculiar velocity (2.5)

$$v_{\mathrm{pec}}(t) = a\sqrt{\widetilde{g}_{ab}\,\frac{dq^a}{dt}\,\frac{dq^b}{dt}}.$$

From (2.31) it follows that

$$\left[\begin{aligned} -c^2 &= g_{\alpha\beta}\,\frac{dq^\alpha}{d\tau}\,\frac{dq^\alpha}{d\tau} = g_{\alpha\beta}\,\frac{dq^\alpha}{dt}\,\frac{dq^\alpha}{dt}\left(\frac{dt}{d\tau}\right)^2 \\ &= \left[-\left(\frac{dq^0}{dt}\right)^2 + a^2\,\widetilde{g}_{ab}\,\frac{dq^a}{dt}\,\frac{dq^b}{dt}\right]\left(\frac{dt}{d\tau}\right)^2 = c^2\left(-1+\frac{v_{\mathrm{pec}}^2}{c^2}\right)\left(\frac{dt}{d\tau}\right)^2. \end{aligned}\right.$$

$$\Longrightarrow 1 = \left(1-\frac{v_{\mathrm{pec}}^2}{c^2}\right)\left(\frac{dt}{d\tau}\right)^2. \tag{2.33}$$

This result shows that $\frac{v_{\mathrm{pec}}^2}{c^2} < 1$, $v_{\mathrm{pec}} < c$. ■

Remark 2.5 Equation (2.33) is equivalent to

$$\boxed{\frac{d\tau}{dt} = \sqrt{1-\frac{v_{\mathrm{pec}}^2}{c^2}}} \tag{2.34}$$

This equation is similar to the well-known formula that in special relativity links proper time and relative time along the history of a massive particle. ●

Remark 2.6 Galaxies, as particles of the cosmic fluid, have zero peculiar velocity so they are sub-luminal particles with $\tau = t$. Consequently, their absolute velocity is

$$V^\alpha \overset{\text{def}}{=} \frac{dq^\alpha}{d\tau} = \frac{dq^\alpha}{dt} = c\,\frac{dq^\alpha}{dq^0} \qquad \begin{cases} V^0 = c, \\ V^a = 0. \end{cases} \tag{2.35}$$

Applying (2.14) we see that the divergence of this vector field is given by

$$\boxed{\nabla_\alpha V^\alpha = 3\,c\,H} \tag{2.36}$$

Chapter 3
Relativistic Cosmic Dynamics

3.1 First Dynamical Postulate: Einstein Equations

Having crossed the relativistic bridge-postulate, our goal is to formulate the physical laws governing the evolution of the scale factor $a(t)$. We know that cosmic spacetime is endowed with the metric (2.30)

$$\boxed{g_{00} = -1, \quad g_{a0} = 0, \quad g_{ab} = a^2(t)\, \widetilde{g}_{ab}(\widetilde{q})}$$

$$\boxed{g_{\alpha\beta}\, dq^\alpha\, dq^\beta = -c^2\, dt^2 + a^2(t)\, \widetilde{g}_{ab}\, dq^a\, dq^b} \tag{3.1}$$

with contravariant components

$$\boxed{g^{00} = -1, \quad g^{a0} = 0, \quad g^{ab} = a^{-2}(t)\, \widetilde{g}^{ab}(\widetilde{q})} \tag{3.2}$$

For now no particular assumptions are made about the reference time $t_\sharp$.

First dynamical postulate. *The space-time metric is determined by Einstein's field equations*

$$\boxed{G^{\alpha\beta} + \Lambda\, g^{\alpha\beta} = \chi\, T^{\alpha\beta}} \tag{3.3}$$

where $\Lambda > 0$ is the **cosmological constant**, *and the constant* χ *is defined by*

$$\boxed{\chi \overset{\text{def}}{=} \frac{8\,\pi\, G_N}{c^4}} \tag{3.4}$$

where G_N is the **Newtonian gravitazional constant**.

S. Benenti, *Mathematical Foundations and Numerical Analysis of the Dynamics of an Isotropic Universe*, https://doi.org/10.1007/978-3-031-64033-9_3

In compliance with the isotropy principle, the components of the tensor $T^{\alpha\beta}$ must be of the type (Theorem 1.8)

$$
T^{\alpha\beta}: \begin{cases} T^{00} = \phi(t) = \text{function of } t \text{ only} \\ T^{a0} = 0 \\ T^{ab} = \psi(t)\, \widetilde{g}^{ab} = \text{function of } t \text{ times } \widetilde{g}^{ab} \end{cases} \tag{3.5}
$$

To make this first postulate applicable, it is necessary to translate Einstein's equations into differential equations in the scale factor $a(t)$. The derivatives with respect to t will be indicated by a superimposed dot.

Theorem 3.1 (i) *The four conservation equations* $\nabla_\alpha T^{\alpha\beta} = 0$ *are equivalent to the single equation*

$$
a\,\dot{\phi} + 3\,\dot{a}\,\left(\phi + a^2\,\psi\right) = 0 \tag{3.6}
$$

which in turns is equivalent to

$$
(\phi\, a^3)^{\dot{}} = -3\, a^4\, \dot{a}\, \psi \tag{3.7}
$$

(ii) *Einstein's ten equations (3.3) are equivalent to the two differential equations*

$$
\frac{\dot{a}^2}{c^2} = \tfrac{1}{3}\, a^2\, (\Lambda + \chi\, \phi) - \widetilde{K} \tag{3.8}
$$

$$
2\,\frac{\ddot{a}}{c^2} = a\left[\tfrac{2}{3}\,\Lambda - \chi\,(\psi\, a^2 + \tfrac{1}{3}\,\phi)\right] \tag{3.9}
$$

(iii) *Equation* (3.9) *is a consequence of* (3.8) *and of the conservation law* (3.6).

Proof (i) Equation (3.6) is the translation of (2.15), item (iii) of Theorem 2.7, in the change from the parameter q^0 to the parameter t, with $\kappa = c$.
(ii) Combining equations (2.29), (3.2) and (3.5) with $\kappa = c$, Einstein's equations (3.3) reduce into a system of two equations

$$
\begin{cases} \dfrac{3}{c^2\, a^2}\,\left(\dot{a}^2 + c^2\, \widetilde{K}\right) - \Lambda = \chi\, \phi, \\[2ex] -\dfrac{1}{c^2\, a^4}\,\left(\dot{a}^2 + 2\, a\, \ddot{a} + c^2\, \widetilde{K}\right) + \dfrac{\Lambda}{a^2} = \chi\, \psi, \end{cases}
$$

that we rewrite in the form

$$
\begin{cases} \dot{a}^2 = c^2 \left[\tfrac{1}{3}\, a^2\, (\Lambda + \chi\, \phi) - \widetilde{K}\right], \\[1ex] 2\, a\, \ddot{a} + \dot{a}^2 + c^2 \widetilde{K} = c^2\, a^2\, \left(\Lambda - \chi\, \psi\, a^2\right). \end{cases}
$$

Let us substitute the first equation into the second one:

$$\Longleftrightarrow \begin{cases} \dot{a}^2 = c^2\left[\frac{1}{3}\,a^2\,(\Lambda+\chi\,\phi) - \widetilde{K}\right] \\ 2\,a\,\ddot{a} + c^2\left[\frac{1}{3}\,a^2\,(\Lambda+\chi\,\phi) - \widetilde{K}\right] + c^2\widetilde{K} = c^2\,a^2\,(\Lambda - \chi\,\psi\,a^2) \end{cases}$$

$$\Longleftrightarrow \begin{cases} \dot{a}^2 = c^2\left[\frac{1}{3}\,a^2\,(\Lambda+\chi\,\phi) - \widetilde{K}\right] \\ 2\,a\,\ddot{a} + \frac{1}{3}\,c^2\,a^2\,(\Lambda+\chi\,\phi) = c^2\,a^2\,(\Lambda - \chi\,\psi\,a^2) \end{cases}$$

$$\Longleftrightarrow \begin{cases} \dot{a}^2 = c^2\left[\frac{1}{3}\,a^2\,(\Lambda+\chi\,\phi) - \widetilde{K}\right] \\ \ddot{a} = \frac{1}{2}\,c^2\,a\left[\frac{2}{3}\,\Lambda - \chi\,\left(\psi\,a^2 + \frac{1}{3}\,\phi\right)\right] \end{cases} \Longleftrightarrow \begin{cases} (3.8) \\ (3.9) \end{cases}$$

(iii) Differentiating of (3.8) we get

$$\frac{2\dot{a}\ddot{a}}{c^2} = \tfrac{2}{3}\,a\dot{a}\,(\Lambda+\chi\,\phi) + \tfrac{1}{3}\,a^2\,\chi\,\dot{\phi} \Longrightarrow \frac{2\ddot{a}}{c^2} = \tfrac{2}{3}\,a\,(\Lambda+\chi\,\phi) + \tfrac{1}{3}\,a^2\,\chi\,\frac{\dot{\phi}}{\dot{a}}.$$

$$(3.6) \Longrightarrow \tfrac{1}{3}\,\frac{a}{\dot{a}}\,\dot{\phi} = -(\phi + a^2\,\psi) \Longrightarrow \frac{2\ddot{a}}{c^2} = \tfrac{2}{3}\,a\,(\Lambda+\chi\,\phi) - a\,\chi\,(\phi + a^2\,\psi)$$

$\Longrightarrow$ (3.9). . ■

> *Remarkable fact: this theorem establishes the relativistic dynamical equations of an isotropic Universe valid for <u>any</u> isotropic energy tensor of the cosmic fluid.*

3.2 Second Dynamical Postulate: Perfect Cosmic Fluid

The following postulate is devoted to specifying the characteristic functions ϕ and ψ of the energy tensor.

> **Second dynamical postulate.** *The cosmic fluid is a* **perfect fluid** *with energy tensor*
>
> $$\boxed{T^{\alpha\beta} = c^{-2}\,(\varepsilon + p)\;V^\alpha\,V^\beta + p\,g^{\alpha\beta}} \qquad (3.10)$$
>
> *where* V^α *is the* **absolute velocity** *of the fluid satisfying the normalization equation*
>
> $$g_{\alpha\beta}\,V^\alpha\,V^\beta = -\,c^2,$$
>
> $\varepsilon(t)$ *is the* **energy density** *and* $p(t)$ *is the* **kinetic pressure**.

The physical dimensions of the elements involved in Einstein equations are given in the following table (see also the tables in Sect. 1.9) (Table 3.1).

Table 3.1 Elements of Einstein equations

Quantity	Dim	Note
Λ	L^{-2}	[a]
$\chi\, T^{\alpha\beta}$	L^{-2}	[a]
$T^{\alpha\beta}$	$\mathsf{M\,L^{-1}\,T^{-2}}$	[b]
χ	$\mathsf{M^{-1}\,L^{-1}\,T^{2}}$	[c]
G_N	$\mathsf{M\,L^{3}\,T^{-2}}$	[d]
p	$\mathsf{M\,L^{-1}\,T^{-2}}$	
ε	$\mathsf{M\,L^{-1}\,T^{-2}}$	

[a] In accordance with the established conventions the coordinates q^α are L-dimensional, so the components of the metric tensor $g_{\alpha\beta}$ and $g^{\alpha\beta}$ are dimensionless and therefore $\mathrm{Dim}\,(R^{\alpha\beta}) = \mathsf{L}^{-2}$. It follows from Einstein's equations that $\mathrm{Dim}\,(\Lambda) = \mathrm{Dim}\,(\chi\, T^{\alpha\beta}) = \mathsf{L}^{-2}$.

[b] Equations (3.14) show that $\mathrm{Dim}\,(T^{00}) = \mathrm{Dim}\,(\varepsilon)$ and $\mathrm{Dim}\,(T^{ab}) = \mathrm{Dim}\,(p)$. It follows from the Table 1.2 that

$$\mathrm{Dim}\,(\varepsilon) = \mathrm{Dim}\,(p) = \mathsf{M\,L^{-1}\,T^{-2}}.$$

[c] $\mathrm{Dim}\,(\chi) = \mathrm{Dim}\,(\chi\, T^{\alpha\beta})/\mathrm{Dim}\,(T^{\alpha\beta}) = \mathsf{L}^{-2}/(\mathsf{M\,L^{-1}\,T^{-2}})$.

[d] $\mathrm{Dim}\,(\chi) = \mathsf{L^{-4}\,T^{4}} \cdot \mathrm{Dim}\,(G_N) \Longrightarrow$
$\mathrm{Dim}\,(G_N) = \mathrm{Dim}\,(\chi)\,\mathsf{L^{4}\,T^{-4}} = \mathsf{M^{-1}\,L^{-1}\,T^{2}\,L^{4}\,T^{-4}} = \mathsf{M^{-1}\,L^{3}\,T^{-2}}$.

Theorem 3.2 *With an energy tensor of the type* (3.10) *the dynamical equations* (3.8), (3.9) *and the conservation equation* (3.6) *become respectively*

$$\boxed{\frac{\dot a^2}{c^2} = \tfrac{1}{3}\, a^2\,(\Lambda + \chi\,\varepsilon) - \widetilde K} \tag{3.11}$$

$$\boxed{\frac{\ddot a}{c^2} = \tfrac{1}{2}\, a\left[\tfrac{2}{3}\,\Lambda - \chi\,(p + \tfrac{1}{3}\,\varepsilon)\right]} \tag{3.12}$$

$$\boxed{a\,\dot\varepsilon + 3\,(\varepsilon + p)\,\dot a = 0} \iff \boxed{(\varepsilon\, a^3)^{\bullet} = -3\,a^2\,\dot a\,p} \tag{3.13}$$

Proof The absolute velocity of the cosmic fluid is given by (2.35). So from the definition (3.10) it follows that

$$\boxed{T^{00} = \varepsilon, \quad T^{a0} = 0, \quad T^{ab} = p\, g^{ab} = p\, a^{-2}\, \widetilde g^{ab}} \tag{3.14}$$

Comparison with (3.5) shows that the characteristic functions are

$$\phi = \varepsilon(t), \quad \psi = a^{-2}(t)\; p(t). \tag{3.15}$$

We then apply Theorem 3.1. ■

Remark 3.1 Equations (3.11)–(3.12) are in the literature referred to or attributed to Friedmann and Lemaître. For a detailed discussion on this topic see Sect. 3.4. •

3.3 Third Dynamical Postulate: State Equations

The three dynamical equations (3.11), (3.12) and (3.13) are equivalent to two independent equations involving three unknown functions: $a(t)$, $\varepsilon(t)$ and $p(t)$. Thus, in order to construct a self-consistent model, we need additional equations representing the physical characteristics of the cosmic fluid.

Third dynamical postulate. (i) *All the components of the cosmic fluid have their own density ε_i which, with their sum, contribute to form a* **total energy density** $\varepsilon(t)$:

$$\boxed{\varepsilon = \sum_i \varepsilon_i} \tag{3.16}$$

(ii) *Each component ε_i generates a pressure p_i proportional to it according to an* **equation of state**

$$\boxed{p_i = w_i\, \varepsilon_i} \tag{3.17}$$

where the dimensionless constant w_i is called the **state parameter** *of component i.*

Remark 3.2 Recall that p_i and ε_i have the same dimension: $\text{Dim}\,(\varepsilon) = \text{Dim}\,(p) = \mathsf{M}\,\mathsf{L}^{-1}\,\mathsf{T}^{-2}$. Thus the constants w_i are dimensionless. •

(iii) *The* **total internal pressure** *is the sum of all pressures*

$$\boxed{p = \sum_i p_i} \tag{3.18}$$

(iv) *Every couple (ε_i, p_i) satisfies the conservation equation* (3.13)

$$\boxed{(\varepsilon_i\, a^3)^{\bullet} + 3\, a^2\, \dot{a}\, p_i = 0}$$

Remark 3.3 All the above equations are supported by arguments reported in astrophysical articles and books. •

Theorem 3.3 *Each component ε_i evolves over time according to the law*

$$\boxed{\varepsilon_i(t) = \frac{\varepsilon_i(t_\natural)}{a^{3(1+w_i)}(t, t_\natural)}} \tag{3.19}$$

Proof $(\varepsilon_i\, a^3)^{\boldsymbol{\cdot}} + 3\, a^2\, \dot a\, p_i = 0 \iff (\varepsilon_i\, a^2 a)^{\boldsymbol{\cdot}} + 3\, a^2\, \dot a\, w_i\, \varepsilon_i = 0$

$$\iff (\varepsilon_i\, a^2)^{\boldsymbol{\cdot}}\, a + \varepsilon_i a^2 \dot a + 3a^2 \dot a w_i \varepsilon_i = 0 \iff (\varepsilon_i a^2)^{\boldsymbol{\cdot}}\, a + \varepsilon_i\, a^2\, \dot a\,(1 + 3w_i) = 0$$

$$\iff \frac{(\varepsilon_i\, a^2)^{\boldsymbol{\cdot}}}{\varepsilon_i\, a^2} + (1 + 3\, w_i)\,\frac{\dot a}{a} = 0$$

$$\iff \log(\varepsilon_i\, a^2) + \log a^{1+3\, w_i} = \log k \quad (k = \text{constant})$$

$$\iff \varepsilon_i\, a^{3(1+w_i)} = k \iff \varepsilon_i(t) = \frac{k}{a^{3(1+w_i)}(t, t_\sharp)}.$$

Value in $t = t_\sharp$: $k = \varepsilon(t_\sharp)$. ■

Remark 3.4 It is a useful exercise to verify that Eq. (3.19) does not depend on the choice of reference time:

$$\varepsilon_i(t) = \frac{\varepsilon_i(t_\sharp)}{a(t, t_\sharp)},\; t = t_\flat \Longrightarrow \varepsilon_i(t_\flat) = \frac{\varepsilon_i(t_\sharp)}{a(t_\flat, t_\sharp)} [\dagger]$$

$$\Longrightarrow \varepsilon_i(t) = \frac{\varepsilon_i(t_\sharp)}{a(t, t_\sharp)} = \frac{\varepsilon_i(t_\sharp)}{a(t, t_\flat)\, a(t_\flat, t_\sharp)} = [\dagger] \frac{\varepsilon_i(t_\flat)}{a(t, t_\flat)}.$$

•

3.4 Comments on Friedmann and Lemaître Equations

Friedman's and Lemaître's equations are widely cited in cosmology texts, where, however, they appear written in various forms, which not only differ in notation but are sometimes not equivalent to each other.

To avoid confusion, as far as Friedmann's equations are concerned, one should refer to the dynamic equations that appear in the original work [10] Uber die Krümmung des Raumes by Friedmann (1922). They are written exactly as follows:

$$\boxed{\begin{cases} (4) \quad \dfrac{R'^2}{R^2} + \dfrac{2\,R\,R''}{R^2} + \dfrac{c^2}{R^2} - \lambda = 0 \\ (5) \quad \dfrac{3\,R'^2}{R^2} + \dfrac{3\,c^2}{R^2} - \lambda = \varkappa c^2 \rho \end{cases}} \qquad \begin{cases} R' = \dfrac{dR}{dx_4}, \\ R'' = \dfrac{d^2R}{dx_4^2}. \end{cases} \tag{3.20}$$

where ρ is stated to be a mass density and $\varkappa$ *eine Konstante*. The coordinate x_4 is time-dimensional and the signature of the metric is $(- - - +)$. These equations come out from Einstein's field equations. The comparison with our Einstein equations. $R^{\alpha\beta} + \left(\Lambda - \frac{1}{2}\, R\right) g^{\alpha\beta} = \chi\, T^{\alpha\beta}$ shows a difference in sign in the second member. This is due to the different signatures of the metric.

Looking at the components of the energy tensor, we observe that
(i) *Friedman takes into account the cosmological constant,*
(ii) *Friedman considers the 'dust' model for galactic fluid* (kinetic pressure p is not present).
(iii) *Friedman considers space curvature positive.*

In our approach, we have seen that the Einstein equations consequent to the energy tensor (3.14) reduce to the differential equations (3.11)–(3.12),

$$\begin{cases} \dfrac{\dot{a}^2}{c^2} = \frac{1}{3}\, a^2\, (\Lambda + \chi\, \varepsilon) - K_\sharp, \\[2ex] \dfrac{\ddot{a}}{c^2} = \frac{1}{2}\, a \left[\frac{2}{3}\, \Lambda - \chi\, (p + \frac{1}{3}\, \varepsilon)\right]. \end{cases} \tag{3.21}$$

Let us compare these equations with Friedman's equations (3.20). To do this we rewrite them in the form

$$\begin{cases} [4] \quad 2\, R\, R'' + R'^2 + c^2 - \lambda\, R^2 = 0, \\ [5] \quad R'^2 + c^2 - \frac{1}{3}\, \left(\lambda + \varkappa\, c^2 \rho\right) R^2 = 0. \end{cases} \tag{3.22}$$

We subtract member to member [4] – [5]:

$$2\, R\, R'' - \lambda\, R^2 + \tfrac{1}{3}\, \left(\lambda + \varkappa\, c^2 \rho\right) R^2 = 0.$$

Since $R \neq 0$, we have $2\, R'' - \frac{2}{3}\, \lambda\, R + \frac{1}{3}\, \varkappa\, c^2 \rho\, R = 0$, that is

$$R'' = \tfrac{1}{6}\, R\, \left(2\, \lambda - \varkappa\, c^2 \rho\right). \tag{3.23}$$

If in (3.23) we put

$$\begin{cases} dx_4 = dt \\ R = a \end{cases} \qquad \begin{cases} R' = \dot{a} \\ R'' = \ddot{a} \end{cases} \tag{3.24}$$

we find equation $\ddot{a} = \frac{1}{6}\, a\, \left(2\, \lambda - \varkappa\, c^2 \rho\right)$ which coincides with the second Eq. (3.21) with $p = 0$,

$$\frac{\ddot{a}}{c^2} = \tfrac{1}{6}\, a\, (2\, \Lambda - \chi\, \varepsilon),$$

provided that $2\, \lambda - \varkappa\, c^2 \rho = c^2\, (2\, \Lambda - \chi\, \varepsilon)$, that is

$$\lambda = c^2\, \Lambda, \quad 3 \varkappa\, \rho = \chi\, \varepsilon. \tag{3.25}$$

In turns, with the substitutions (3.24), the second equation [5] in (3.22) becomes

$$\dot{a}^2 = \tfrac{1}{3}\, \left(\lambda + \varkappa\, c^2 \rho\right) a^2 - c^2.$$

Because (3.25), $\dot{a}^2 = \frac{1}{3}\, c^2\, a^2\, (\Lambda + \chi\, \varepsilon) - c^2$, this equation coincide with our first equation (3.21)

$$\frac{\dot{a}^2}{c^2} = \tfrac{1}{3}\, a^2\, (\Lambda + \chi\, \varepsilon) - K_\sharp$$

provided that $K_\sharp = 1$. This proves item (iii).

As far as Lemaître equations are concerned, the main reference should be the article [12] *Un Univers homogène de masse constante et de rayon croissant rendant compte de la vitesse radiale des nébuleuses extra-galactiques* (1927). The gravitational field equations are there presented (without demonstration) in the form

$$\boxed{\begin{cases} (2)\;\; 3\,\dfrac{R'^2}{R^2} + \dfrac{3}{R^2} = \lambda + \kappa\,\rho \\[2ex] (3)\;\; 2\dfrac{R''}{R} + \dfrac{R'^2}{R^2} + \dfrac{1}{R^2} = \lambda - \kappa\,\rho \end{cases}} \qquad {}' = \frac{d}{dt}, \tag{3.26}$$

where λ is the cosmological constant, κ is the '*Einstein constant*', ρ is the density of the '*total energy*' and R is '*le rayon de l'espace*'.

A third equation is then introduced

$$\boxed{(4)\;\; \frac{d\rho}{dt} + 3\,\frac{R'}{R}\,(\rho + p) = 0} \tag{3.27}$$

where p is the '*density of radiant energy*'. This equation is claimed (without demonstration) to be equivalent to the four momentum energy tensor conservation equations. It is in full agreement with the conservation law (3.13) $a\,\dot{\varepsilon} + 3\,(\varepsilon + p)\,\dot{a} = 0$ by setting $\rho = \varepsilon$ and $R = a$.

Therefore, the '*rayon de l'espace*' R of Lemaître is actually the scale factor a, which is a dimensionless quantity. It follows that the first members of the Lemaître equations (2) and (3) are dimensionally incongruent, so it becomes difficult to examine their relationships to our dynamical equations (3.11)–(3.12). Indeed, many of the equations written by Lemaître suffer from this inconsistency, mainly because of the absence of the speed of light, probably set equal to 1. An example is given by the equation of a light ray

$$\boxed{(20)\;\; \sigma_2 - \sigma_1 = \int_{t_1}^{t_2} \frac{dt}{R}}$$

where the first member has the dimension of a length because it is understood to be the difference of two positions in space, while the second member is a time (if R is dimensionless). This equation is the archetype of our Eq. (5.2), Sect. 5.2,

$$\boxed{d_{AB}(t_\sharp) = c \int_{t_{eA}}^{t_{rB}} \frac{dt}{a(t, t_\sharp)}}$$

which gives the distance $d_{AB}(t_\sharp)$ of two cosmic bodies A and B in the reference space $S_{t_\sharp}$, with t_{eA} the time of emission from A of a light ray and t_{rB} the time of reception from B.

3.5 Fourth Dynamical Postulate: Matter and Radiation

From the third postulate and Theorem 3.3, we can begin a path toward the construction of **many-component models of the Universe**, already attempted by some authors but without clear results. With a fourth postulate, we restrict our interest to a simpler **two-component model**: **matter** and **radiation**.

Fourth dynamical postulate. (i) *widespread in the Universe there are two fundamental energy densities:*

$$\boxed{\varepsilon = \varepsilon_m + \varepsilon_r} \tag{3.28}$$

With two distinct characteristic properties:

(ii) *the* **matter density** ε_m *that does not generate pressure,*

$$\boxed{p_m(t) = 0, \quad \text{i.e. } w_m = 0} \tag{3.29}$$

(iii) *the* **radiation density** $\varepsilon_r(t)$ *that generates pressure,*

$$\boxed{p_r(t) = \frac{1}{3}\,\varepsilon_r(t) \quad \text{i.e. } w_r = \frac{1}{3}} \tag{3.30}$$

(iv) *The matter density is in turn the sum*

$$\boxed{\varepsilon_m = \varepsilon_b + \varepsilon_c} \tag{3.31}$$

of a **baryonic energy density** ε_b *and a* **cold dark matter energy density** ε_c.

(v) *There exists a date* t_{eq} *of* **matter-radiation equilibrium** *in which the two densities* ε_m and ε_r *have equal value*

$$\boxed{\varepsilon_m(t_{\text{eq}}) = \varepsilon_r(t_{\text{eq}})} \tag{3.32}$$

We call the cosmological model based on this dynamical postulate the **matter-radiation model** *(MR-model). The existence of a time t_{eq} of matter-radiation equilibrium will play a crucial role.*

Remark 3.5 As already claimed in Remark 3.3 all the above equations are supported by arguments reported in astrophysical articles and books. •

Remark 3.6 In addition to the energies of mass and radiation, cosmologists consider the cosmological constant Λ to be representative of a third type of energy, the **dark energy**, which generates negative pressure within the cosmic fluid. We do not adopt this interpretation in our approach because, in compliance with our postulates, Λ is a **universal constant** while the densities of matter and radiation energies depend on time. •

By virtue of Eq. (3.19) the densities ε_m and ε_r evolve in time according to the laws

$$\boxed{\varepsilon_m(t) = \frac{\varepsilon_{m\sharp}}{a^3(t, t_\sharp)}} \tag{3.33}$$

$$\boxed{\varepsilon_r(t) = \frac{\varepsilon_{r\sharp}}{a^4(t, t_\sharp)}} \tag{3.34}$$

Consequently, the equality between matter density and radiation density (3.32) turns out to be equivalent to

$$\boxed{\varepsilon_{r\sharp} = a(t_{\text{eq}}, t_\sharp)\, \varepsilon_{m\sharp}} \tag{3.35}$$

Furthermore, by virtue of the Eqs. (3.11), (3.28), (3.33) and (3.34) the scale factor $a(t, t_\sharp)$ referred to a generic time $t_\sharp$ is governed by the dynamical equation

$$\boxed{\frac{\dot{a}^2(t, t_\sharp)}{c^2} = \tfrac{1}{3}a^2(t, t_\sharp)\left[\Lambda + \frac{\chi\, \varepsilon_{m\sharp}}{a^3(t, t_\sharp)} + \frac{\chi\, \varepsilon_{r\sharp}}{a^4(t, t_\sharp)}\right] - K_\sharp} \tag{3.36}$$

Dividing by a^2, this equation becomes

$$\boxed{\frac{H^2(t)}{c^2} = \frac{1}{3}\left[\Lambda + \frac{\chi\, \varepsilon_{m\sharp}}{a^3(t, t_\sharp)} + \frac{\chi\, \varepsilon_{r\sharp}}{a^4(t, t_\sharp)}\right] - \frac{K_\sharp}{a^2(t, t_\sharp)}} \tag{3.37}$$

For $t = t_\sharp$ we have the normalization condition $a(t_\sharp, t_\sharp) = 1$ and we obtain the spatial curvature $K_\sharp = K(t_\sharp)$ at the reference time expressed in terms of the densities evaluated in $t_\sharp$:

$$\boxed{K_\sharp = \frac{1}{3}\left[\Lambda + \chi\,(\varepsilon_{m\sharp} + \varepsilon_{r\sharp})\right] - \frac{H_\sharp^2}{c^2}} \tag{3.38}$$

3.6 Dynamical Equations of the MR-Model

So far the scale factor has been referred to a generic time $t_\sharp$. However, we should note that the estimates of the various cosmological entities are resulting from measurement made in the present epoch only, and that it is from these estimates that we have to infer the evolution of the Universe in the 'past' as well as in the 'future'. Hard task indeed.

Therefore, from now on we will take the **present time** t_0 as the reference time.

Astronomers are nowadays able to measure the following **cosmological parameters** denoted by Ω_*:

$$\boxed{\text{dark energy}: \quad \Omega_\Lambda \stackrel{\text{def}}{=} \tfrac{1}{3}\,\frac{c^2}{H_0^2}\,\Lambda \iff \Lambda = \frac{3\,H_0^2}{c^2}\,\Omega_\Lambda} \tag{3.39}$$

$$\boxed{\text{matter}: \quad \Omega_m \stackrel{\text{def}}{=} \tfrac{1}{3}\,\frac{c^2}{H_0^2}\,\chi\,\varepsilon_{m0} \iff \chi\,\varepsilon_{m0} = \frac{3\,H_0^2}{c^2}\,\Omega_m} \tag{3.40}$$

$$\boxed{\text{radiation}: \quad \Omega_r \stackrel{\text{def}}{=} \tfrac{1}{3}\,\frac{c^2}{H_0^2}\,\chi\,\varepsilon_{r0} \iff \chi\,\varepsilon_{r0} = \frac{3\,H_0^2}{c^2}\,\Omega_r} \tag{3.41}$$

$$\boxed{\text{spatial curvature}: \quad \Omega_K \stackrel{\text{def}}{=} \Omega_\Lambda - \Omega_m - 1} \tag{3.42}$$

All these parameters are *dimensionless*.[1]

Theorem 3.4 *In the MR-model, the evolution of the scale factor $a(t, t_0)$ is governed by equation*

$$\boxed{\dot{a}^2 = H_0^2\left[\Omega_\Lambda\,a^2 + \Omega_m\,a^{-1} + \Omega_r\,a^{-2}\right] - c^2\,K_0} \tag{3.43}$$

where H_0 and K_0 are the present day values of the Hubble factor and of the spatial curvature, which are related to each other by equation

$$\boxed{K_0 = \frac{H_0^2}{c^2}\,(\Omega_\Lambda + \Omega_m + \Omega_r - 1)} \tag{3.44}$$

Remark 3.7 By virtue of (3.44) the dynamical equation (3.43) takes the alternative form

$$\boxed{\dot{a}^2 = H_0^2\left[1 + \Omega_\Lambda\,(a^2 - 1) + \Omega_m\,(a^{-1} - 1) + \Omega_r\,(a^{-2} - 1)\right]} \tag{3.45}$$

[1] In [4], p.37, Ω_K is defined with opposite sign.

This form has the advantage that it does not involve directly the spatial curvature K_0, which, according to current measurements, is so 'small' that it is even believed to be null. •

Proof Substitute $\Omega_\Lambda \stackrel{\text{def}}{=} \frac{1}{3}\,\frac{c^2}{H_0^2}\,\Lambda,\ \Omega_m \stackrel{\text{def}}{=} \frac{1}{3}\,\frac{c^2}{H_0^2}\,\chi\,\varepsilon_{m0},\ \Omega_r \stackrel{\text{def}}{=} \frac{1}{3}\,\frac{c^2}{H_0^2}\,\chi\,\varepsilon_{r0}$ in (3.43):

$$\left[\begin{array}{l} \dot a^2 = H_0^2\left[\Omega_\Lambda\, a^2 + \Omega_m\, a^{-1} + \Omega_r\, a^{-2}\right] - c^2\, K_0 \\ = H_0^2\left[\frac{1}{3}\,\frac{c^2}{H_0^2}\,\Lambda\, a^2 + \frac{1}{3}\,\frac{c^2}{H_0^2}\,\chi\,\varepsilon_{m0}\, a^{-1} + \frac{1}{3}\,\frac{c^2}{H_0^2}\,\chi\,\varepsilon_{r0}\, a^{-2}\right] - c^2\, K_0 \\ = \frac{1}{3}\, c^2\left[\Lambda\, a^2 + \chi\,\varepsilon_{m0}\, a^{-1} + \chi\,\varepsilon_{r0}\, a^{-2}\right] - c^2\, K_0 \end{array}\right.$$

$$\Longrightarrow\quad \frac{\dot a^2(t, t_0)}{c^2} = \frac{1}{3}\left[\Lambda\, a^2 + \chi\,\varepsilon_{m0}\, a^{-1} + \chi\,\varepsilon_{r0}\, a^{-2}\right] - K_0.$$

Then compare this result with Eq. 3.36 re-written for $t_\sharp = t_0$. ■

3.7 Relationship Between Matter and Radiation Densities

The existence of a time t_{eq} in which the densities of matter and radiation have equal value—fourth dynamical postulate, item (v)—plays a crucial role in the theoretical and numerical analysis of the MR-model.

Theorem 3.5 *The values of the cosmological parameters Ω_r and Ω_m are related by the equation*

$$\boxed{\Omega_r = a_{\text{eq}}\,\Omega_m = \frac{\Omega_m}{1 + z_{\text{eq}}}} \tag{3.46}$$

where

$$\boxed{a_{\text{eq}} \stackrel{\text{def}}{=} a(t_{\text{eq}}, t_0) = \frac{1}{1 + z_{\text{eq}}}} \tag{3.47}$$

is the scale factor at time t_{eq} and z_{eq} the corresponding redshift.

Proof As will be seen in Sect. 6.1, the redshift z is related to time t (interpreted as **emission time**) by Eq. (6.7)

$$\frac{1}{a(t, t_0)} = 1 + z. \tag{3.48}$$

By virtue of (3.40) and (3.41),

$$\Omega_m = \frac{1}{3}\,\frac{c^2}{H_0^2}\,\chi\,\varepsilon_{m0}, \quad \Omega_r = \frac{1}{3}\,\frac{c^2}{H_0^2}\,\chi\,\varepsilon_{r0},$$

it follows that

$$\frac{\Omega_r}{\Omega_m} = \frac{\varepsilon_r(t_0)}{\varepsilon_m(t_0)} = a_{\text{eq}}.$$

■

A remarkable consequence of Eq. (3.46) is:

Theorem 3.6 *In the MR-model the evolution of the scale factor with reference time t_0 is governed by the dynamical equation*

$$\boxed{\dot{a}^2 = H_0^2 \left[1 + \Omega_\Lambda (a^2 - 1) + \Omega_m \left(a^{-1} - 1 + \frac{a^{-2} - 1}{1 + z_{\text{eq}}} \right) \right]} \tag{3.49}$$

where only the **four primary data** *are involved:*

$$\begin{cases} H_0 & \textit{present time value of the Hubble factor,} \\ \Omega_\Lambda & \textit{dark energy parameter,} \\ \Omega_m & \textit{matter energy parameter,} \\ z_{\text{eq}} & \textit{redshift corresponding to time } t_{\text{eq}}. \end{cases} \tag{3.50}$$

In fact, thanks to (3.46) the dynamical equation (3.45) is transformed into (3.49).

As mentioned in the Preface, towards the end of the 1990s an agreement was found called **concordance cosmology** with which the Λ**CDM – Lambda Cold Dark Matter model** was founded, briefly called the **standard model**, whose dynamical equations

$$\boxed{\dot{a}^2 = H_0^2 \left[\Omega_\Lambda \, a^{2-3(1+w)} + (\Omega_b + \Omega_c)\, a^{-1} + \Omega_r \, a^{-2} + \Omega_K \right]} \tag{3.51}$$

involve <u>five</u> constants:

$$\begin{cases} w & \text{parameter of the dark energy equation of state,} \\ \Omega_b & \text{baryon density parameter,} \\ \Omega_c & \text{cold dark matter parameter,} \\ \Omega_r & \text{radiation density parameter,} \\ \Omega_K & \text{curvature parameter.} \end{cases}$$

By setting

$$\begin{cases} w = -1 \\ \Omega_b + \Omega_c = \Omega_m \\ H_0^2 \Omega_K = -c^2 K_0 \end{cases} \tag{3.52}$$

we find Eq. (3.43) and fall back to the MR-model. However,

> The MR-model cannot be considered as a special case of the standard model for the following reasons:
>
> 1. The MR-model is based on postulates clearly expressed in mathematical terms.
> 2. The MR-model requires knowledge of only four primary data (3.50).
> 3. In the MR-model the spatial curvature is positive (Theorem 3.7, Sect. 4.2).
> 4. The estimates of other relevant cosmological entities that are deducible from the four primary data are in excellent agreement with those obtained from the most recent observational data.
> 5. The scale factor curve $a(t)$, whose analytical and numerical expression will be given in Sects. 4.7 and 4.9, collimates excellently with the profile exposed by Riess in his Nobel lecture [20].

3.8 The Problem of the Spatial Curvature

According to the most recent observational data, the curvature parameter (3.42) Ω_K turns out to have such a small value that one is inclined to conclude that *the Universe is flat*. Such a conclusion, however, may be erroneous. In fact we know that in a dynamical equation to put a *small parameter* equal to *zero* can completely change the behavior of the solutions.

An illuminating example of this misunderstanding is given by the historical debate about the cosmological constant Λ. Because its value is in fact very small, for years and years it was felt that it could be neglected and thus omitted from Einstein's equations.

Such an omission, however, is an operation incompatible with the very principles of general relativity that imply the presence of this constant.

Finally, it has only recently been realized that, even if of very small value, Λ causes over long timescales an acceleration of the scale factor growth. In Sect. 4.13 of this chapter, we examine the relevant behavioral differences between models with $\Lambda \neq 0$ and those with $\Lambda = 0$.

In any case, a first answer to the problem of the spatial curvature in the MR-model is given by the following theorem:

Theorem 3.7 *In the MR-model, the spatial curvature cannot be zero.*

Proof **Part 1**. The cosmological parameters defined in (3.40) and (3.41) are measured at the present time. They are defined in a similar way when measured at a generic time t, i.e.

$$\Omega_m(t) \stackrel{\text{def}}{=} \tfrac{1}{3}\,\frac{c^2}{H^2(t)}\,\chi\,\varepsilon_m(t), \quad \Omega_r(t) \stackrel{\text{def}}{=} \tfrac{1}{3}\,\frac{c^2}{H^2(t)}\,\chi\,\varepsilon_r(t).$$

On the other hand, choosing t_0 as the reference time, Eqs. (3.33) and (3.34) become

$$\varepsilon_m(t) = \frac{\varepsilon_m(t_0)}{a^3(t)}, \quad \varepsilon_r(t) = \frac{\varepsilon_r(t_0)}{a^4(t)},$$

so that

$$\Omega_m(t) = \tfrac{1}{3}\,\frac{c^2}{H^2(t)}\,\chi\,\frac{\varepsilon_m(t_0)}{a^3(t)}, \quad \Omega_r(t) = \tfrac{1}{3}\,\frac{c^2}{H^2(t)}\,\chi\,\frac{\varepsilon_r(t_0)}{a^4(t)}.$$

Since, see again (3.40) and (3.41),

$$\chi\,\varepsilon_{m0} = \frac{3\,H_0^2}{c^2}\,\Omega_m, \quad \chi\,\varepsilon_{r0} = \frac{3\,H_0^2}{c^2}\,\Omega_r,$$

it follows that

$$\Omega_m(t) = \frac{H_0^2}{H^2(t)}\,\frac{1}{a^3(t)}\,\Omega_m, \quad \Omega_r(t) = \frac{H_0^2}{H^2(t)}\,\frac{1}{a^4(t)}\,\Omega_r,$$

i.e.

$$\Omega_m(t) + \Omega_r(t) = \frac{H_0^2}{H^2(t)}\left(\frac{\Omega_m(t_0)}{a^3(t,t_0)} + \frac{\Omega_r(t_0)}{a^4(t,t_0)}\right). \tag{3.53}$$

Part 2. Let by hypothesis be $K_0 = 0$. Then, by virtue of equation (3.44) we have

$$\Omega_\Lambda + \Omega_m + \Omega_r = 1.$$

In this equation, the cosmological parameters Ω_m and Ω_r are referred to the present epoch t_0, so it should be written

$$\Omega_\Lambda + \Omega_m(t_0) + \Omega_r(t_0) = 1. \tag{3.54}$$

We cannot write $\Omega_\Lambda(t_0)$ because Λ is by postulate a *universal constant* independent of t. It should also be noted that *this equation must be independent of the scale factor, i.e. the type of evolution of the Universe*. Moreover, by the sign permanence theorem of the curvature (Sect. 1.5) this is zero at all times. Therefore Eq. (3.54) is any time valid. It follows that the sum

$$\Omega_m(t) + \Omega_r(t) = 1 - \Omega_\Lambda$$

does not depend on time, whatever the evolution of the Universe. However, this is contrary to the conclusion of Part 1 of the proof, that is Eq. (3.53), which shows the time dependence of this sum. Thus, the hypothesis $K_0 = 0$ leads to an absurd. ■

Remark 3.8 The proof of this theorem does not need the numerical estimates of the cosmological parameters Ω_*. •

3.9 Weierstrass Equation

A **Weierstrass equation** is a first-order differential equation of the type[2]

$$\boxed{\dot{a}^2 = W(a)} \tag{3.55}$$

where the function $W(a)$ is called **Weierstrass function**.

The dynamical equation of the MR-model is a Weierstrass equation. The function W may take the three different forms (3.43), (3.45) and (3.49):

$$\boxed{\begin{aligned} W(a) &= H_0^2\left[\Omega_\Lambda\, a^2 + \Omega_m\, a^{-1} + \Omega_r\, a^{-2}\right] - c^2 K_0 \\ W(a) &= H_0^2\left[1 + \Omega_\Lambda\,(a^2 - 1) + \Omega_m\,(a^{-1} - 1) + \Omega_r\,(a^{-2} - 1)\right] \\ W(a) &= H_0^2\left[1 + \Omega_\Lambda\,(a^2 - 1) + \Omega_m\left(a^{-1} - 1 + \frac{a^{-2} - 1}{1 + z_{\mathrm{eq}}}\right)\right] \end{aligned}} \tag{3.56}$$

The peculiarity of a W-equation is that, even if we are not able to solve it, we can infer the main properties of the solutions from the analysis of the function $W(a)$ i.e. its graph in the plane $x = a$, $y = \dot{a}^2$. To do this we must interpret a solution $a(t)$ as the motion of a virtual point on the a-axis. Since the first member of the equation (3.55) is never negative, this motion can take place only in the intervals of the a-axis where $W(a)$ is non-negative.

These intervals are bounded by the **zeros** of $W(a)$, solutions of the equation $W(a) = 0$, that is, by the points where the graph of W touches or crosses the a-axis. Note that a zero a_* of $W(a)$ is a **stopping point** because $W(a_*) = 0$ is equivalent to $\dot{a} = 0$ for $a = a_*$. Therefore, the search for any zeros has a priority character.

A point a_* where $W(a_*) = 0$ and $W'(a_*) \neq 0$ is said to be a **simple zero**.[3] A zero where instead it is $W'(a_*) = 0$ is said to be **multiple zero**. These two types of zeros have quite different properties.[4]

---

[2] See [13] and *Weierstrass collected works*, V.II.

[3] The superscript $'$ denotes the derivative with respect to a.

[4] More details can be found in Sect. 3.12.

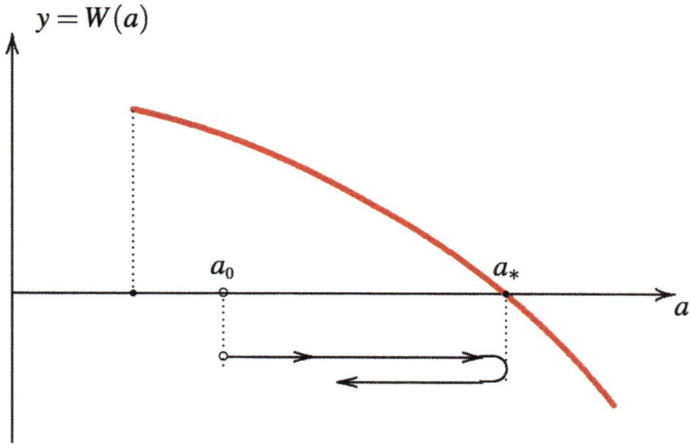

Fig. 3.1 Simple zero a_* as inversion point

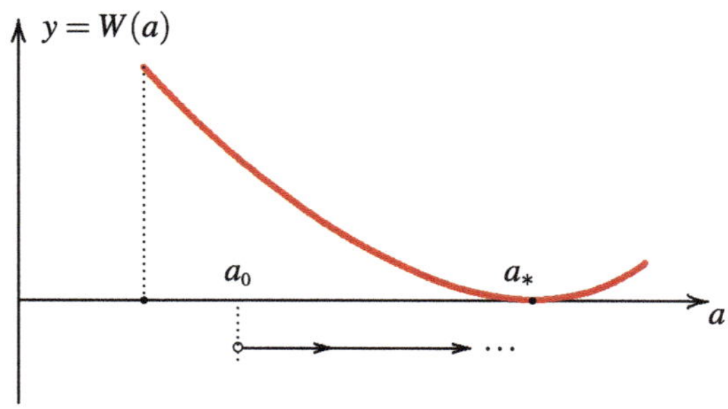

Fig. 3.2 Multiple zero as asymptotic objective

[1] A simple zero a_* is an **inversion point** in the sense that *if the virtual point moves toward a simple zero a_* then reaches it in a finite* <u>*time*</u>, *there it stops and then moves off again in the opposite direction* (see Fig. 3.1).

[2] A multiple zero is an **asymptotic objective** in the sense that *if the virtual point moves toward a multiple zero a_*, then it never stops and never reaches it* (see Fig. 3.2).

[3] Assume that:

$$
\begin{aligned}
&\text{(i) } W(a) \text{ is positive in a closed interval } [a_1, a_2];\\
&\text{(ii) the moving point is located in } a_1 \textit{ at the time } t = t_1;\\
&\text{(iii) } \dot{a}(t_1) > 0, \text{ the point is moving toward } a_2.
\end{aligned}
\tag{3.57}
$$

Then the velocity $\dot{a}(t)$ remains positive for $t > t_1$ and the W-equation equation becomes equivalent to

$$dt = \frac{da}{\sqrt{W(a)}}.$$

Consequently, the position a_2 is reached at time

$$t_2 = t_1 + \int_{a_1}^{a_2} \frac{dx}{\sqrt{W(x)}} \tag{3.58}$$

and the integral

$$\boxed{t_2 - t_1 = \int_{a_1}^{a_2} \frac{dx}{\sqrt{W(x)}}} \tag{3.59}$$

provides the time taken in the trip from a_1 to a_2.

3.10 Profiles of the Universe

By **profile of the Universe** we mean the graph of the scale factor $a(t)$ on the coordinate plane (t, a). The following properties apply to profiles.

(i) By virtue of the normalization condition $a(t_\natural, t_\natural) = 1$, and a profile of $a(t, t_\natural)$ with reference time $t_\natural$ passes through the point $(t_\natural, 1)$.

(ii) Two profiles that differ by a translation along the t axis have to be considered equivalent.

(iii) As it is explicitly said in Theorem 3.4, the profiles corresponding to the functions $W(a)$ (3.56) have t_0 as reference time, so they pass through the point $(t_0, 1)$.

(iv) If a profile starts from the origin $t = 0$ with $a = 0$ then, by posing $t_1 = 0$ and $a_1 = 0$ in (3.59), then the integral

$$\boxed{t(a) = \int_0^a \frac{dx}{\sqrt{W(x)}}} \tag{3.60}$$

provides the elapsed time in the transition from the initial state of the Universe to the state where the scale factor has an assigned value a. Consequently, if t_0 (today) is the reference time, the inverse function $t(a)$ of (3.60) is the profile of the solution $a(t, t_0)$ of the W-equation, so that, placing $a = 1$ in the integral (3.60) we get the **age of the Universe**:

$$\boxed{t_0 = \int_0^1 \frac{dx}{\sqrt{W(x)}}} \tag{3.61}$$

Table 3.2 Key-events

a_*	Event	t_*
1	Today state	t_0
a_{eq}	Balance of matter and radiation density	t_{eq}
a_q	Beginning of accelerated expansion	t_q
a_{re}	Reionization (beginningg of light emission)	t_{re}

(v) A Universe profile may pass through points (t_*, a_*) of particular interest, which we call **key-events**. If we place $a = a_*$ in the integral (3.60)

$$\boxed{t_* = \int_0^{a_*} \frac{dx}{\sqrt{W(x)}}} \tag{3.62}$$

then we obtain the date t_* of this key-event.

(vi) In the MR-model the profile of the Universe has four key-events (Table 3.2):

3.11 Qualitative Profiles of the MR-Model

The basic analytical properties of the Weierstrass function (3.43) are:

$$\lim_{a\to 0} W(a) = +\infty, \quad \lim_{a\to +\infty} W(a) = +\infty \tag{3.63}$$

$$W'(a) = H_0^2 \left[2\,\Omega_\Lambda\, a - \Omega_m\, a^{-2} - 2\,\Omega_r\, a^{-3}\right] \tag{3.64}$$

$$W''(a) = H_0^2 \left[2\,\Omega_\Lambda + 2\,\Omega_m\, a^{-3} + 6\,\Omega_r\, a^{-4}\right] \tag{3.65}$$

The second derivative is everywhere positive so $W(a)$ is a convex function. Since the function $W(a)$ tends to $+\infty$ for $a \to 0$ and for $a \to +\infty$, it has a unique point of minimum $a_q \neq 0$, i.e. the root of the fourth-degree equation $W'(a) = 0$

$$\boxed{\Omega_\Lambda\, a^4 - \tfrac{1}{2}\,\Omega_m\, a - \Omega_r = 0} \tag{3.66}$$

which is inferred from (3.64). Even if we do not know the values of a_q and $W_q = W(a_q)$ we can still plot a summary qualitative graph of $W(a)$ (see Fig. 3.3).

Differentiating the equation $\dot{a}^2 = W(a)$ we find equation $2\,\dot{a} = W'(a)\,\dot{a}$; which in turns, for $\dot{a} \neq 0$, provides the equation

$$2\,\ddot{a} = W'(a). \tag{3.67}$$

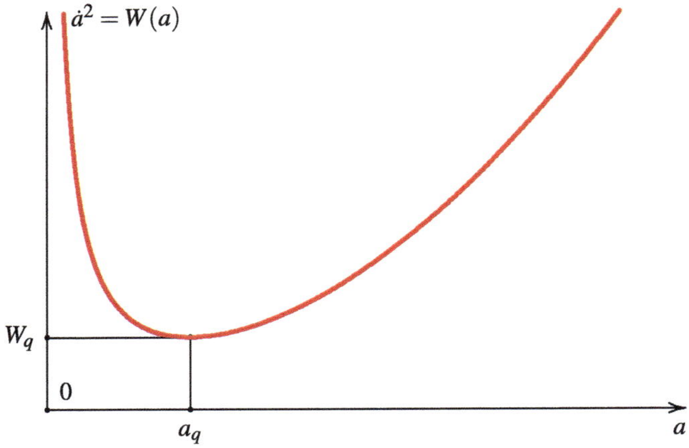

Fig. 3.3 Summary graph of the Weierstrass function (3.43)

This equation shows that the acceleration $\ddot{a}$ of the scale factor cancels in a root of $W'(a)$. For various reasons, cosmologists have introduced the **deceleration parameter**

$$q \stackrel{\text{def}}{=} -\frac{a\,\ddot{a}}{\dot{a}^2} \tag{3.68}$$

which cancels out when $\ddot{a} = 0$. This justifies the notation a_q for the root of $W'(a)$.

In studying the graph of $W(a)$ we have to consider two cases: $W_q > 0$ and $W_q < 0$.[5] The case $W_q > 0$ is illustrated at the top of Fig. 3.4.

Since $W(a)$ is everywhere positive, there are no zeros, no stopping points. Consequently $a(t)$ is always increasing (expansion) or always decreasing (contraction). Let us consider the first case only. The point of minimum marks the transition between a slowed expansion and an accelerated expansion (middle part of the figure). The qualitative profile is shown below where at a_q there is an **inflection point**, corresponding to a time t_q. Note that, by virtue of (3.56), $W_q > 0$ is satisfied for $K_0 \leq 0$.

The case $W_q < 0$ is illustrated in Fig. 3.5. There are two simple zeros (inversion points) between which $W(a)$ is negative. Consequently, we have two separate possible profiles outside the zeros, each with two different orientations. On the left we have a **pseudo-cyclic profile** with a **big-bang** and a **big-crunch**. On the right we have a profile, viable in two directions, with unbounded $a(t)$.

[5] The case $W_q = 0$ is unrealistic because it would correspond to a predetermined relationship between the cosmological parameters expressed by the Eq. (3.66).

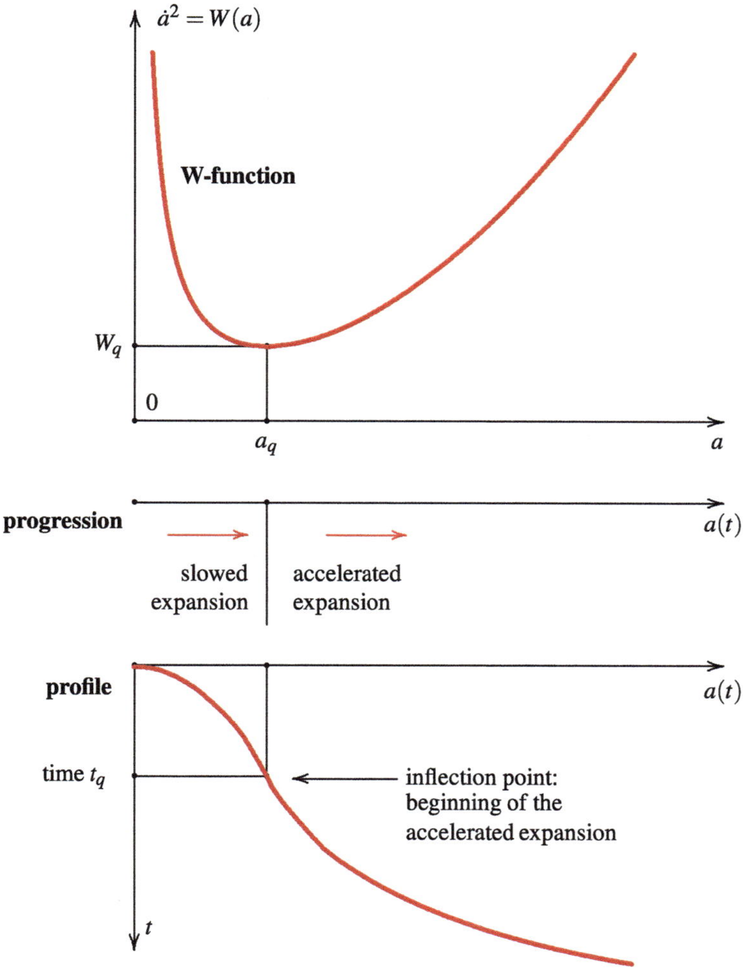

Fig. 3.4 Qualitative profile of the Universe in the case $W_q > 0$

3.12 Complements on the Weierstrass Equation

[1] A simple zero a_* is an **inversion point**.

Proof Let us consider the case of Fig. 3.1: $a_0 < a_*$. The time taken in the path from the starting position a_0 to the arrival position a_* is given by the improper integral

$$t_\star - t_\sharp = \int_{a_0}^{a_*} \frac{da}{\sqrt{W(a)}} = \lim_{a \to a_*} \int_{a_0}^{a} \frac{da}{\sqrt{W(a)}}. \tag{3.69}$$

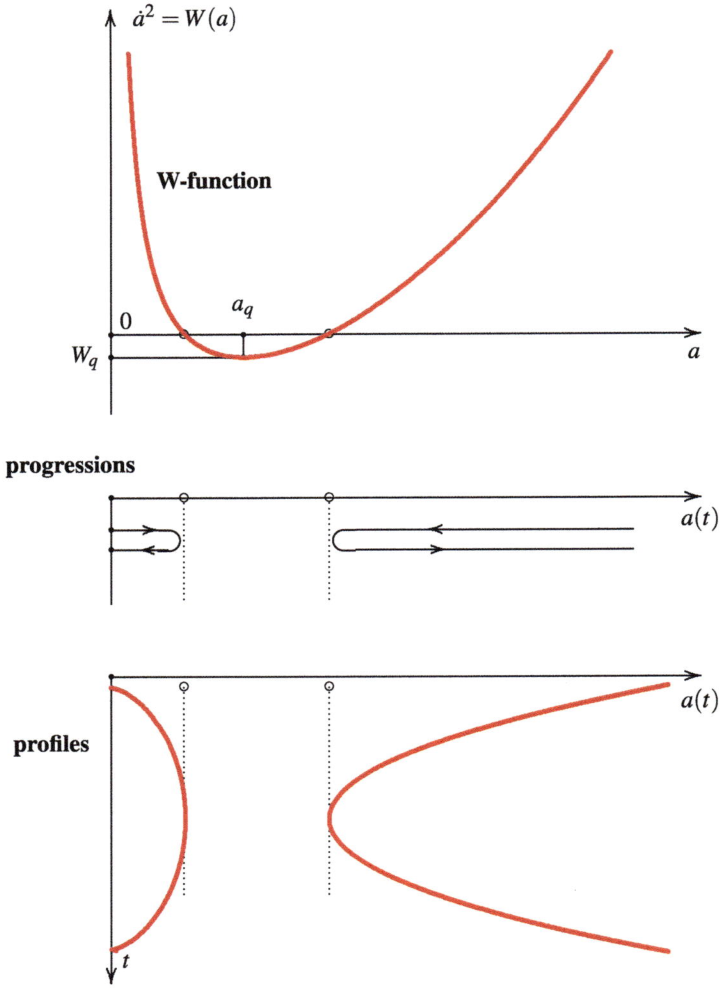

Fig. 3.5 Qualitative profile of the Universe for $W_q < 0$

Since the integrand function is positive for $a < a_*$, the integral from a_0 to a is an increasing function of a. Thus two possibilities arise: (i) the limit (3.69) is finite or (ii) it is $+\infty$. Let us consider the Taylor expansion of W in the left neighborhood of a_*:

$$W(a) = (a - a_*)\, W'(a_*) + \tfrac{1}{2}\,(a - a_*)^2\, W''(a_*) + \ldots$$

For a simple zero $W'(a_*) \neq 0$. In our case $W'(a_*) < 0$, so we can write

$$W(a) = (a - a_*)\, W'(a_*) \left[1 + \tfrac{1}{2}\,(a - a_*)\, \frac{W''(a_*)}{W'(a_*)} + \ldots \right] = f(a)\, g(a) \qquad (3.70)$$

where, to the left of a_*, $f(a) \stackrel{\text{def}}{=} (a - a_*)\, W'(a_*)$ is a positive function, while

$$g(a) \stackrel{\text{def}}{=} 1 + \tfrac{1}{2}\,(a - a_*)\,\frac{W''(a_*)}{W'(a_*)} + \ldots$$

is bounded and positive because $\lim\limits_{a \to a_*} g(a) = 1$. Let us now apply one of the fundamental theorems concerning improper integrals:

Theorem 3.8 (i) *The improper integral*

$$\int_{x_0}^{x_*} \frac{dx}{(x_1 - x)^p}, \quad p > 0, \tag{3.71}$$

is convergent if $p < 1$, *divergent if* $p \geq 1$. (ii) *The same holds for the integral*

$$\int_{x_0}^{x_*} \frac{h(x)}{(x_* - x)^p}\, dx, \quad p > 0, \tag{3.72}$$

where $h(x)$ *is an integrable function such that* $\lim_{x \to x_*} h(x) \neq 0$.

Proof (i)

$$I(x) \stackrel{\text{def}}{=} \int_{x_0}^{x} \frac{dz}{(a_* - z)^p} =$$
$$\begin{cases} p = 1: & -\Big[\log(a_* - z)\Big]_{x_0}^{x} = \log(x_* - x_0) - \log(x_* - x). \\ p \neq 1: & -\frac{1}{1-p}\Big[(x_* - z)^{1-p}\Big]_{x_0}^{x} = \frac{1}{1-p}\Big[(x_* - x_0)^{1-p} - (x_* - x)^{1-p}\Big]. \end{cases}$$

$$\begin{cases} p = 1: & \lim\limits_{x \to x_*} I(x) = +\infty. \\ p < 1: & \lim\limits_{x \to x_*} I(x) = \frac{1}{1-p}\,(a_* - a_0)^{1-p}. \\ p > 1: & \lim\limits_{x \to x_*} I(x) = \frac{1}{1-p}\left[(x_* - x_0)^{1-p} - \lim\limits_{x \to x_*} (x_* - x)^{1-p}\right] = -\infty. \end{cases}$$

(ii) is a corollary of (i). ∎

The integral (3.69) is within this theorem for

$$p = \tfrac{1}{2}, \quad h(a) = \sqrt{|W'(a_*)|\, g(a)}.$$

So it is convergent. The derivative of the Weierstrass equation (3.55) yields the equation $2\,\dot{a} = W'(a)\,\dot{a}$. For $\dot{a} \neq 0$ this equation is equivalent to

$$2\,\ddot{a} = W'(a). \tag{3.73}$$

However, for reasons of continuity, this equation is also valid in the limit $\dot{a} \to 0$, so (3.73) holds for every $\dot{a}$. Since $W'(a_*) \neq 0$, equation (3.73) implies $\ddot{a} \neq 0$. Consequently, when the moving point reaches a_* it immediately moves away from it because at a_* its acceleration is not zero. But it cannot go beyond a_* where W becomes negative. So it is forced to turn back. ■

[2] A multiple zero is an **asymptotic objective**.

Proof Let us put ourselves in the case of Fig. 3.2: the point moves toward a zero a_* starting from $a_0 < a_*$. It cannot stop before a_* because $W > 0$ in the whole interval $a_0 \leq a < a_*$. The arrival time in a_* is defined as in the (3.69). But now it is $W'(a_*) = 0$ and Taylor's development in the surroundings of a_* becomes

$$W(a) = \tfrac{1}{2}\,(a - a_*)^2\, W''(a_*) + \tfrac{1}{6}\,(a - a_*)^3\, W'''(a_*) + \ldots$$

However, we cannot exclude that it is $W''(a_*) = 0$, $W''(a_*) = 0$, etc. Therefore, if $q \geq 2$ is the order of the first nonzero derivative in a_*, we can write

$$W(a) = \tfrac{1}{q!}\,(a - a_*)^q\, W^{(q)}(a_*) + \tfrac{1}{(q+1)!}\,(a - a_*)^{q+1}\, W^{(q+1)}(a_*) + \ldots$$

With similar reasoning as in the previous demonstration concerning a simple zero, we come to apply the theorem on improper integrals stated above with $p = \frac{1}{2}\,q \geq 1$, In this case the integral (3.69) is divergent. ■

[3] **Unbounded Weierstrass function.**

Let $W(a) > 0$ in an interval $a_0 \leq a < a_*$ such that $\lim_{a \to a_*} W(a) = +\infty$. If $\dot{a}(t_0) > 0$ then the moving point reaches a_* in a finite time $t_\star$ determined by the ordinary integral

$$t_\star - t_0 = \int_{a_0}^{a_*} \frac{da}{\sqrt{W(a)}}.$$

Indeed, we have $\lim_{a \to a_*} 1/\sqrt{W(a)} = 0$ and the integrand function can be extended by continuity to a_* where it takes the value 0 (Fig. 3.6).

[4] **Positive W-function in an unbounded interval**

$$a_0 \leq a < +\infty.$$

If the point moves toward $+\infty$ then it can reach $+\infty$ in a finite time t_∞ as long as the improper integral

$$t_\infty - t_0 = \int_{a_0}^{+\infty} \frac{da}{\sqrt{W(a)}} \overset{\text{def}}{=} \lim_{a \to +\infty} \int_{a_0}^{a} \frac{dz}{\sqrt{W(z)}} \tag{3.74}$$

is convergent. It is clear that this happens if $W(a)$ has sufficiently high growth (Fig. 3.7).

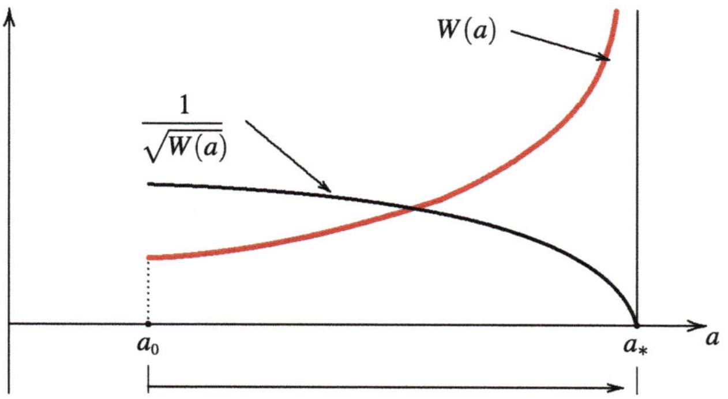

Fig. 3.6 Vertical asymptote in a_*

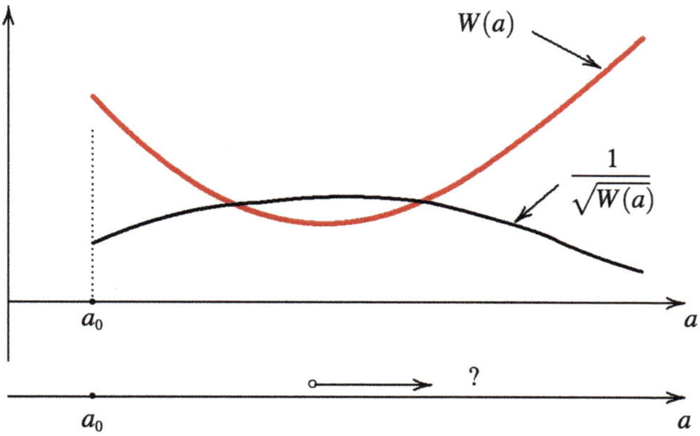

Fig. 3.7 W-function defined in an unbounded interval $[a_0, +\infty)$

5 A Weierstrass equation (3.55) is integrable by separation of variables, being equivalent to the two equations

$$\frac{da}{\sqrt{W(a)}} = \pm dt. \tag{3.75}$$

If $I(a)$ is any integral function of the first member then

$$I(a) = \pm t + \text{arbitrary constant.}$$

Thus, if $a = I^{-1}(y)$ is the inverse function of $y = I(a)$, then in an interval where $I(a)$ is increasing the functions

$$a_{\pm}(t) = I^{-1}(\pm t + \text{constant})$$

are two solutions of the equation. On the other hand, we can take as an arbitrary constant a prefixed value $t_\star$ of time, so that we can write the two solutions in the form

$$a_\pm(t) = \pm I^{-1}(t - t_\star).$$

It follows that $a_+(t)$ is an increasing function of $t - t_\star$ while $a_-(t)$ is decreasing. These solutions, which we call **dual**, are obtained from each other by an inversion of time t and finally translated along the t-axis by the difference $t - t_\star$.

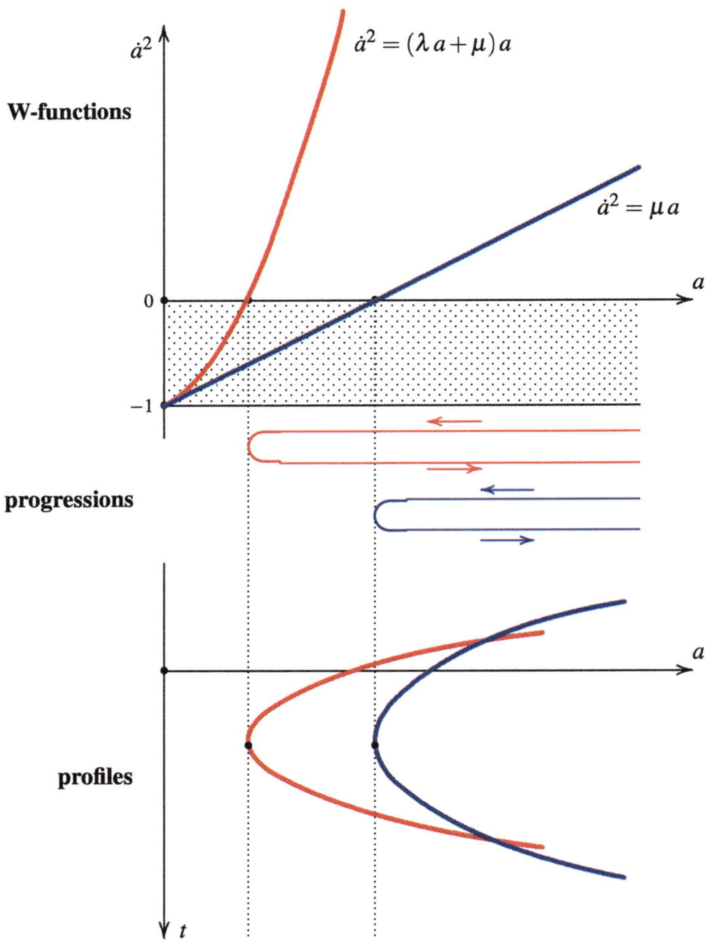

Fig. 3.8 Linear and quadratic W-functions with simple zero

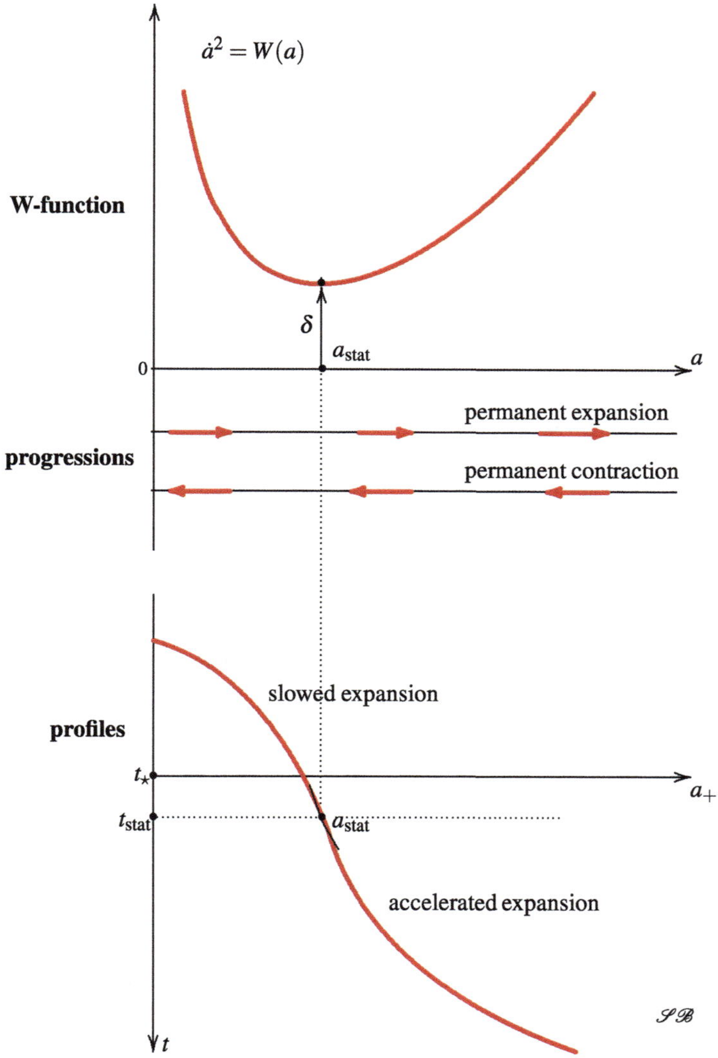

Fig. 3.9 Convex W-function unbounded in $a = 0$ and with a positive minimum

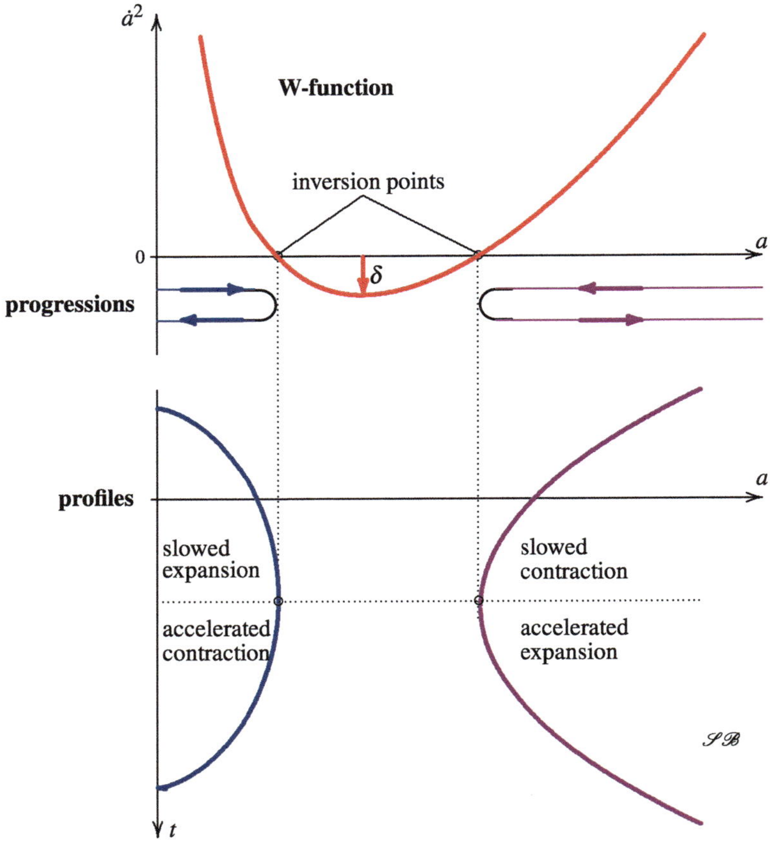

Fig. 3.10 Convex W-function unbounded in $a = 0$ with a negative minimum

Proposed here, without comment, are some examples of how from the graph of the W-function $W(a)$ we can deduce the salient qualitative properties of the solutions of the W-equation $\dot{a}^2 = W(a)$. The following figures show:

(i) At the top, the graphs of $W(a)$ in the plane $(a, \dot{a}^2)$.

(ii) In the middle, the type of progression of the moving point on the a-axis.

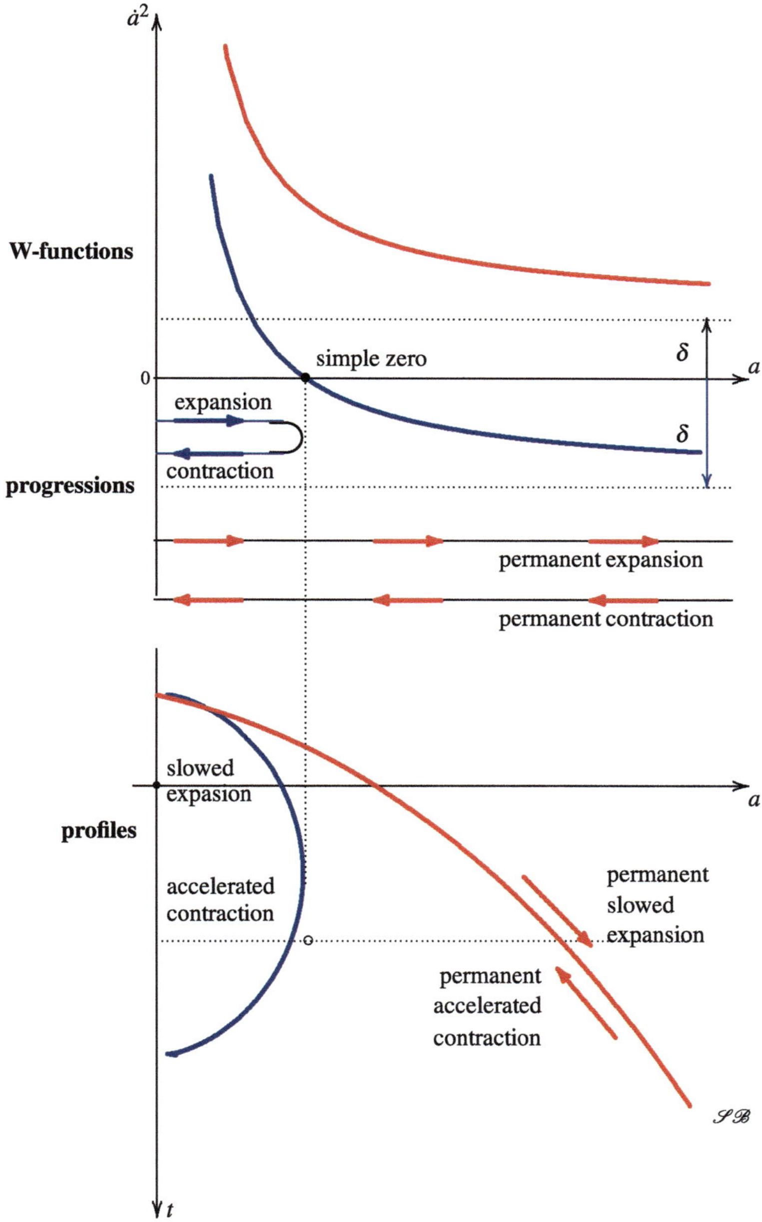

Fig. 3.11 Decreasing W-functions with limit δ at $+\infty$. For $\delta > 0$ no root, for $\delta < 0$ a simple root

(iii) In the lower part, the profiles of the solutions in the (t, a) plane (Figs. 3.8, 3.9, 3.10 and 3.11).

Chapter 4
Numerical Cosmology

4.1 Gathering Cosmological Data

To put the MR-model at work, numerical estimates of the four primary data (3.50) are needed. A large number of *data reports* have been published in recent years. Special attention has been paid to the Planck project reports published in Astronomy & Astrophysics. The Table 4.1 shows some estimates provided by the most recent reports [1–3].

Additional primary data estimates were taken from the 2010 Wilkinson Microwave Anisotropy Probe project [21]:

	From [3] T.8
H_0	66.93 ± 0.62
Ω_Λ	missing
Ω_m	0.3202 ± 0.0087
z_{eq}	missing
t_0	13.826 ± 0.025
z_{re}	8.24 ± 0.88

Finally, the most recent and reliable Hubble factor estimate obtained in conjunction with the gravitational wave detection in August 2017 was taken into account [14]:

The original version of the chapter has been revised: Incorrect equations have been updated. A correction to this chapter can be found at https://doi.org/10.1007/978-3-031-64033-9_8

S. Benenti, *Mathematical Foundations and Numerical Analysis of the Dynamics of an Isotropic Universe*, https://doi.org/10.1007/978-3-031-64033-9_4

Table 4.1 From A&A Planck reports (2016)

	[1] T.21 first column	[1] T.21 second column	[2] T.8
H_0	67.31 ± 0.96	67.27 ± 0.66	$\boxed{67.74 \pm 0.46}$
Ω_Λ	0.685 ± 0.013	0.6844 ± 0.0091	$\boxed{0.6911 \pm 0.0062}$
Ω_m	0.315 ± 0.013	0.3156 ± 0.0091	$\boxed{0.3089 \pm 0.0062}$
z_{eq}	3393 ± 49	3395 ± 33	$\boxed{3371 \pm 23}$
t_0	13.813 ± 0.038	13.813 ± 0.026	$\boxed{13.799 \pm 0.021}$
z_{re}	$9.9^{+1.8}_{-1.6}$	$10.0^{+1.7}_{-1.5}$	$\boxed{8.8^{+1.2}_{-1.1}}$

Table 4.2 From WMAP 7th year (2010)

H_0	$70.01 \pm 1.3\ km\,s^{-1}Mpc^{-1}$
Ω_Λ	0.721 ± 0.015
Ω_{tot}	1.0052 ± 0.0064
t_0	13.75 ± 0.11

$$\boxed{H_0 = 70.0^{+12.0}_{-8.0}\ km\,s^{-1}Mpc^{-1}} \tag{4.1}$$

This estimate is in full agreement with that in Table 4.2.

Remark 4.1 The value of H_0 strongly influences the calculation of many cosmological quantities, such as, for example, the age t_0 of the Universe. It is therefore appropriate to take into account both the estimate (4.1), denoted by $\widehat{H}_0$, and the estimate given in the third column [2]T.8 of Table 4.1 (Planck), denoted by $\bar{H}_0$. *

In summary, we will base the numerical analysis of the MR-model on the following primary data values:

Regarding the uncertainty margins of the values of the cosmological parameters Ω_Λ and Ω_m in Table 4.1 we observe that

$$\Omega_\Lambda = \begin{bmatrix} 0.6911 + 0.0062 \\ 0.6911 \\ 0.6911 - 0.0062 \end{bmatrix} = \begin{bmatrix} 0.6973 \\ 0.6911 \\ 0.6849 \end{bmatrix}$$

Table 4.3 Primary data of the MR-model

$\widehat{H}_0$	70.0 $km\, s^{-1} Mpc^{-1}$	Ligo [14]
$\bar{H}_0$	67.74 $km\, s^{-1} Mpc^{-1}$	Planck [2]
Ω_Λ	0.6911	Planck [2]
Ω_m	0.3089	Planck [2]
z_{eq}	3371	Planck [2]

$$\Omega_m = \begin{bmatrix} 0.3089 + 0.0062 \\ 0.3089 \\ 0.3089 - 0.0062 \end{bmatrix} = \begin{bmatrix} 0.3151 \\ 0.3089 \\ 0.3027 \end{bmatrix}$$

$$\Omega_\Lambda + \Omega_m = \begin{bmatrix} 0.6973 \\ 0.6911 \\ 0.6849 \end{bmatrix} + \begin{bmatrix} 0.3151 \\ 0.3089 \\ 0.3027 \end{bmatrix} = \begin{bmatrix} 1.0124 \\ 1 \\ 0.9876 \end{bmatrix} \tag{4.2}$$

4.2 In the MR-Model the Spatial Curvature Is Positive

In Sect. 3.8 it was shown that, regardless of the numerical estimates of the cosmological parameters, the spatial curvature in the MR-model cannot be zero. Owning the above estimates, we can now state that:

Theorem 4.1 *In the MR-model, the spatial curvature is positive.*

Proof For the mean values of the cosmological parameters Ω_Λ and Ω_m the equality

$$\Omega_\Lambda + \Omega_m = 1. \tag{4.3}$$

holds. By virtue of (3.39) and (3.40) this equality translates into

$$\tfrac{1}{3}\,\frac{c^2}{H_0^2}\,\Lambda + \tfrac{1}{3}\,\frac{c^2}{H_0^2}\,\chi\,\varepsilon_{m0} = 1$$

i.e.

$$\frac{c^2}{H_0^2}\left[\Lambda + \chi\,\varepsilon_{m0}\right] = 3. \tag{4.4}$$

Such a precise numerical relationship between the universal constants Λ, c, χ, and the quantities measured in the present epoch, H_0 and ε_{m0}, is rather doubtful. However, a reasonable interpretation of this fact[1] is the following: *the sum* $\Omega_\Lambda + \Omega_m$ lit is equal to a number different from 1 but contained in the uncertainty interval (4.2)

$$0.9876 < \Omega_\Lambda + \Omega_m < 1.0124.$$

In light of these facts, it should be noted that if in (3.44) we place $\Omega_\Lambda + \Omega_m = 1$ we find

$$\boxed{K_0 = \frac{H_0^2}{c^2}\,\Omega_r} \tag{4.5}$$

Therefore, as the number $\Omega_\Lambda + \Omega_m$ approaches the number 1 the value of the curvature gets closer and closer to the *positive* value given by (4.5). □

Indeed, as is to be expected, the curvature will turn out to be extremely small (see Sect. 4.5).

4.3 Closed Universe Model

We resume here the discussion just outlined in Sect. 1.5 on the topology of spatial sections. The simplest topology that a three-dimensional manifold with constant positive curvature can assume is that of the sphere $\mathbb{S}_3$.[2]

In this case each spatial section S_t is diffeomorphic to a sphere $\mathbb{S}_3$ of radius $r(t)$ represented by equation

$$x_1^2 + x_2^2 + x_3^2 + x_4^2 = r^2(t)$$

in the tetra-dimensional space $\mathbb{R}^4$ with orthonormal coordinates (x_1, x_2, x_3, x_4), centered at the origin. Such a cosmological model is called **closed Universe model**. It can be shown (see Sect. 7.1) that the curvature of each S_t is

$$K(t) = \frac{1}{r^2(t)}. \tag{4.6}$$

The radius of the hypersphere $\mathbb{S}_3 = S_t$ contracts and expands with time while the center O remains fixed. In this way we obtain a **radial representation** of the evolution of the Universe. Figure 4.1 provides a two-dimensional view. Each galaxy A is a point moving along a line exiting from the origin. At each instant t two galaxies A and B stand on the hypersphere of radius $r(t)$ and are separated by an arc

[1] This is one of the so-called **coincidence problems** that arise in cosmology.

[2] In obedience to our **principle of simplicity** we can assume this as a (last) Postulate.

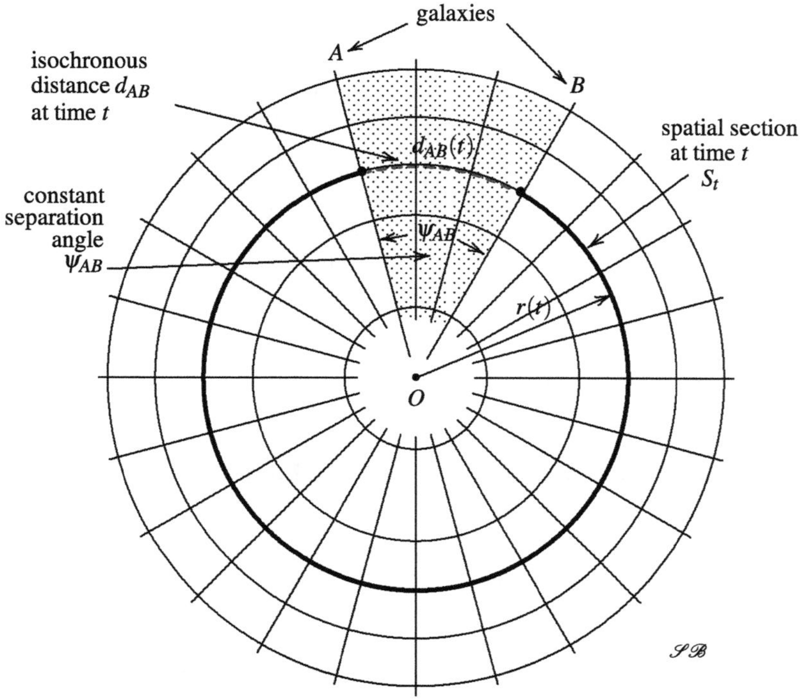

Fig. 4.1 Radial representation at time t of a closed Universe

of maximum circle (geodesic arc) whose length is equal to the isochronous distance $d_{AB}(t)$. The straight lines joining A and B to the center form an angle ψ_{AB} such that

$$\boxed{d_{AB}(t) = \psi_{AB}\, r(t)} \tag{4.7}$$

We call it the **separation angle** of the two galaxies. This angle remains constant over time. The maximum distance between two galaxies is $\pi\, r$, half the length $2\,\pi\, r$ of a maximum circle. It follows that the maximum angular separation is $\psi_{\max} = \pi$. Figure 4.1 represents a snapshot at time t of a closed Universe. Figure 4.2 shows the movie formed by the sequence of these snapshots. This movie is immersed in the affine space $\mathbb{R}^5$. The individual 'frames' are affine spaces $\mathbb{R}^4$. The circles are the hyperspheres of radius $r(t)$ representing the spatial sections.

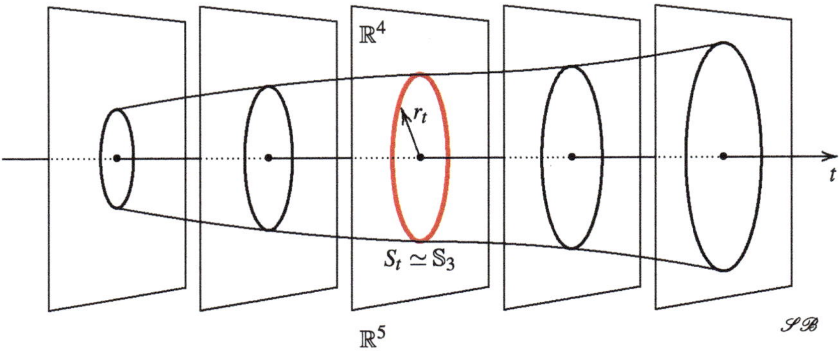

Fig. 4.2 Penta-dimensional affine space-time of a closed Universe

4.4 Measurement Units of the Hubble Factor

The physical dimension of H_0 is T^{-1}. Astronomers measure H_0 in units $km\ s^{-1} Mpc^{-1}$. However, for our purposes it is more convenient to use Gyr (Giga-year, billion years) as the unit of time and then, for homogeneity, to use $Glyr$ (Giga-lightyear, billion light-years) as the unit of length.

Despite the fact that these units are even deemed 'deplorable' by some 'insiders',[3] they are instead more perceptible to the uninitiated people and moreover their use facilitates considerably the numerical treatment of cosmological models in general. For example, with this choice the application of formula (4.5) for the calculation of today's spatial curvature does not require the intervention of the numerical value of the speed of light c, which with this convention results to be equal to 1.

The transition from $km\ s^{-1} Mpc^{-1}$ to Gyr^{-1} is obtained by applying two successive conversion rules.

(i) Conversion from *megaparsecs* to *kilometers*:[4]

$$\boxed{1\ Mpc = 3.0856776 * 10^{19}\ km} \qquad \boxed{\frac{km}{Mpc} = \frac{1}{3.0856776} * 10^{-19}} \tag{4.8}$$

(ii) Conversion from *seconds* to *years*:

$$\begin{gathered}\boxed{1\ s = 3.1709791983765 * 10^{-8}\ yr} \\ \boxed{\frac{yr}{s} = \frac{1}{3.1709791983765} * 10^{8}}\end{gathered} \tag{4.9}$$

---

[3] See [19], Table 2.1.

[4] From [6].

Combining (i) and (ii) we find

$$1\ km\, s^{-1} Mpc^{-1} = \frac{10^{-19}}{3.0856776}\ s^{-1} = \frac{10^{-11}}{3.0856776 * 3.17097919...}\ yr^{-1}$$
$$= \frac{10^{-2}}{3.0856776 * 3.17097919...}\ Gyr^{-1} \quad \Longrightarrow$$

$$\boxed{1\ km\ s^{-1} Mpc^{-1} \simeq 0.0010220121532...\ Gyr^{-1}} \tag{4.10}$$

Then for the two estimates of H_0 in Table 4.3 we find:

$$\boxed{\widehat{H}_0 = 70.00\ km\ s^{-1} Mpc^{-1} \simeq 0.0715408\ Gyr^{-1}} \tag{4.11}$$

$$\boxed{\bar{H}_0 = 67.74\ km\ s^{-1} Mpc^{-1} \simeq 0.0692311\ Gyr^{-1}} \tag{4.12}$$

4.5 Space Curvature Estimates

By applying formula (4.5) with $\Omega_r = 0.3089$ (Table 4.3) and the two estimates (4.11) and (4.12) of H_0, we find

$$\widehat{H}_0 \mapsto \boxed{K_0 = \frac{\widehat{H}_0^2}{c^2}\ \Omega_r = 0.46885499... * 10^{-6}\ Glyr^{-2}} \tag{4.13}$$

$$\bar{H}_0 \mapsto \boxed{K_0 = \frac{\bar{H}_0^2}{c^2}\ \Omega_r = 0.43906907... * 10^{-6}\ Glyr^{-2}} \tag{4.14}$$

It follows that in the two cases the radius of curvature $r_0 = \dfrac{1}{\sqrt{K_0}}$ is

$$\widehat{H}_0 \mapsto \boxed{r_0 = 1460.4299...\ Glyr} \tag{4.15}$$

$$\bar{H}_0 \mapsto \boxed{r_0 = 1509.1540...\ Glyr} \tag{4.16}$$

The lengths of spatial geodesics (Remark 7.2) are

$$\widehat{H}_0 \mapsto \boxed{\ell_{\max} = 2\,\pi\, r_0 = 9176.1519...\ Glyr} \tag{4.17}$$

$$\bar{H}_0 \mapsto \boxed{\ell_{\max} = 9482.2946...\ Glyr} \tag{4.18}$$

4.6 Weierstrass Functions of the MR-Model

By placing $\Omega_\Lambda + \Omega_m = 1$ in the third expression (3.56) the Weierstrass function becomes

$$\boxed{W(a) = H_0^2 \left[a^2 + \Omega_m \left(a^{-1} - a^2 + \frac{a^{-2} - 1}{1 + z_{\text{eq}}} \right) \right]} \tag{4.19}$$

Because of what was said in Sect. 4.2, equality $\Omega_\Lambda + \Omega_m = 1$ is only 'approximate'. However, this approximation agrees with the tolerance of the estimates given in the tables of Sect. 4.1.[5]

Taking into account the estimates from the Table 4.3 and of the conversion rules (4.11) and (4.12) we have two numerical expressions of $W(a)$ in Gyr^{-2} units:

$$\widehat{H}_0 \mapsto \boxed{W(a) = (0.0715408)^2 \left[a^2 + 0.3089 * \left(\frac{1}{a} - a^2 + \frac{a^{-2} - 1}{3372} \right) \right]} \tag{4.20}$$

$$\bar{H}_0 \mapsto \boxed{W(a) = (0.0692311)^2 \left[a^2 + 0.3089 * \left(\frac{1}{a} - a^2 + \frac{a^{-2} - 1}{3372} \right) \right]} \tag{4.21}$$

The corresponding W-functions are plotted in Fig. 4.3.

4.7 Pointwise Numerical Profiles of the MR-Model

We are now in a position to determine numerically the two profiles of the Universe corresponding to the two Weierstrass functions (4.20) and (4.21). It is first necessary to go back to what was said in 3.11 about describing the profiles of the Universe within the MR model, which can be obtained by virtue of the advantages offered by a Weierstrass equation. It should therefore be noted that the $W(a)$ graphs in Fig. 4.3 are quite similar to the one plotted in Fig. 3.4 in the above paragraph. The function $W(a)$ in this figure corresponds to a profile $a(t)$ starting from the origin of the plane (t, a), that is, satisfying the initial condition $a(0) = 0$.

This result, obtained at the qualitative level, is of the utmost importance at the numerical level because it empowers us to apply what was said in (iv) of 3.10, that is, to use the integral (3.60) to calculate a profile satisfying the above initial condition. Indeed, by calculating the integral (3.60)

---

[5] This 'agreement' will be substantiated by the numerical results obtained in the following.

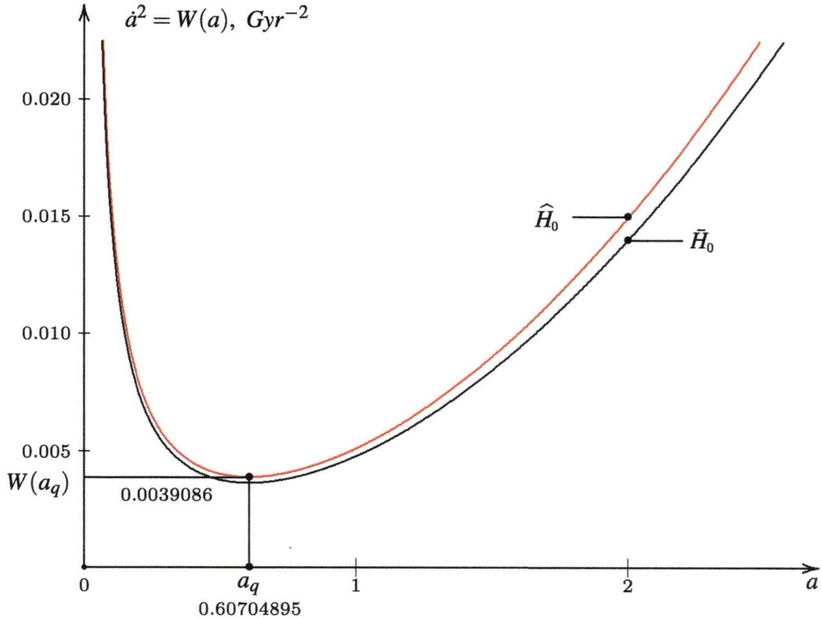

Fig. 4.3 Weierstrass functions of the MR-model for different values of H_0

$$t(a) = \int_0^a \frac{dx}{\sqrt{W(x)}}$$

for a sufficiently fine sequence of values of a one obtains a numerical pointwise representation of the function $t(a)$, whose inverse $a(t)$ provides a pointwise numerical representation of the profile of the MR-Universe. It is clear that, the more dense the sequence of a is, the more the points of the graphs of $t(a)$ and $a(t)$ tend to be indistinguishable. But the real great virtue of this method is that the small but inevitable error in the computation of one $t(a)$ does not affect the computation of the next $t(a)$, as it does in neat step-by-step integration methods.

The profiles of the MR-Universe obtained by this method, corresponding to the two values $\widehat{H}_0$ and $\bar{H}_0$ of the Hubble 'constant' are plotted in Fig. 4.4. As noted in item (iv) of 3.11, in both cases the reference time of the scale factor is the present time t_0.

These profiles are in perfect agreement with Fig. 4.5 taken from A.G. Riess' Nobel Lecture and conveniently reworked.[6] The envelope of the spatial sections (red curve) has the same trend as the scale factor.

[6] 2011 Nobel Prize in Physics, together with Saul Perlmutter and Brian P. Schmidt '*for the discovery of the accelerating expansion of the Universe through observations of distant supernovae*'.

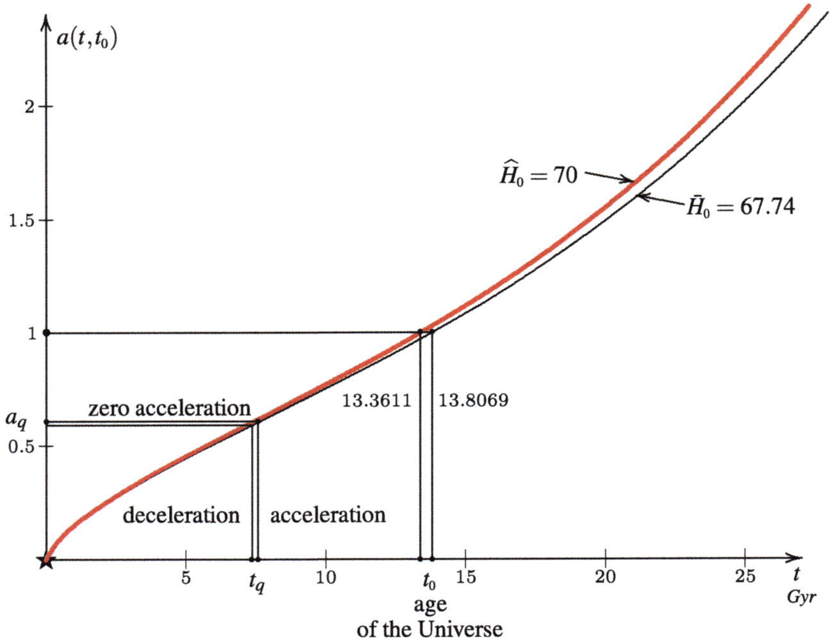

Fig. 4.4 Pointwise numerical profiles of the scale factor in the MR-model with different H_0

Fig. 4.5 Riess profile

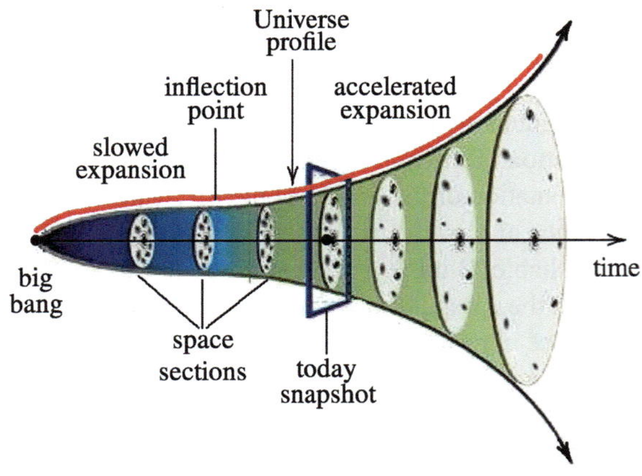

4.8 Age of the Universe and Other Key-Dates

Let's recall here the Table 3.2 of key-events:

a_*	$key - event$	t_*
1	today's state	t_0
a_{eq}	equal matter and radiation densities	t_{eq}
a_q	zero acceleration of the scale factor	t_q
a_{re}	reionization	t_{re}

(4.22)

Given a value of the key-scale factor a_*, the corresponding date is calculated by the integral (3.62)

$$\boxed{t_* = \int_0^{a_*} \frac{dx}{\sqrt{W(x)}}} \tag{4.23}$$

where the numerical expressions of the two W-fuctions corresponding to the estimates $\widehat{H}_0$ and $\bar{H}_0$ of H_0 are given in (4.20) and (4.21):

$$\widehat{H}_0 \mapsto \boxed{W(a) = (0.0715408)^2 \left[a^2 + 0.3089 * \left(\frac{1}{a} - a^2 + \frac{a^{-2} - 1}{3372}\right)\right]}$$

$$\bar{H}_0 \mapsto \boxed{W(a) = (0.0692311)^2 \left[a^2 + 0.3089 * \left(\frac{1}{a} - a^2 + \frac{a^{-2} - 1}{3372}\right)\right]}$$

* **Age of the Universe** t_0.[7] Scale factor $a_0 = 1$. The time t_0 is given by the integral

$$t_0 = \int_0^1 \frac{dx}{\sqrt{W(x)}}$$

which provides the two estimates

$$\boxed{\begin{array}{l} \widehat{H}_0 \mapsto t_0 \simeq 13.36116 \ \ Gyr \\ \bar{H}_0 \mapsto t_0 \simeq 13.80692 \ \ Gyr \end{array}} \tag{4.24}$$

* **Date t_{eq} of equal matter and radiation density**. Recalling the estimate $z_{eq=3371}$ from Table 4.3 and the definition (3.47) of a_{eq} results in

[7] The beginning of the Universe is at $t = 0$.

$$\boxed{a_{\rm eq} \stackrel{\rm def}{=} \frac{1}{1+z_{\rm eq}} = 0.2965599... * 10^{-3}} \tag{4.25}$$

* **Date t_q of scale factor zero acceleration**. The key-scale factor a_q, minimum value of the function $W(a)$, is the positive root of the Eq. (3.66)

$$\Omega_\Lambda\, a_q^4 - \tfrac{1}{2}\, \Omega_m\, a_q - \Omega_r = 0$$

can now be rewritten as

$$\Omega_\Lambda\, a_q^4 - \Omega_m \left(\tfrac{1}{2}\, a_q + a_{\rm eq}\right) = 0.$$

The resulting estimates are

$$\boxed{a_q \simeq 0.60704} \tag{4.26}$$

$$\boxed{\widehat{H}_0 \mapsto t_q \simeq 7.37949\ Gyr, \quad \bar{H}_0 \mapsto t_q \simeq 7.62569\ Gyr} \tag{4.27}$$

* **Beginning $t_{\rm re}$ of the reionization epoch** (*Let there be light*). From the estimate $z_{\rm re} = 8.8$ of the reionization redshift in the third column [2] T.8 of Table 4.1 we derive the corresponding value of the scale factor

$$\boxed{a_{\rm re} \stackrel{\rm def}{=} \frac{1}{1+z_{\rm re}} \simeq 0.10204} \tag{4.28}$$

from which we get the dates

$$\boxed{\widehat{H}_0 \mapsto t_{\rm re} \simeq 0.54409\ Gyr, \quad \bar{H}_0 \mapsto t_{\rm re} \simeq 0.56224\ Gyr} \tag{4.29}$$

Remark 4.2 From (4.28) we derive the value of the radiation density Ω_r by means of the (3.46):

$$\boxed{\Omega_r = a_{\rm eq}\, \Omega_m = 0.91607354... * 10^{-4}} \tag{4.30}$$

Note its small value in respect to Ω_m. This confirms the well-known fact in the present epoch that matter predominates. *

Remark 4.3 The decrease of H_0 in the transition from $\widehat{H}_0$ to $\bar{H}_0$ has the effect of increasing all the dates, especially the age of the Universe (4.24). In turn, the increase in $t\,(a_*)$ dating has the effect of shifting the profile of the Universe toward the future, as shown in Fig. 4.4. *

Remark 4.4 Observe that the value $t_0 = 13.8069\ Gyr$ corresponding to

$$\bar{H}_0 = 67.74\ km\ s^{-1} M!pc^{-1}$$

is in full agreement with the value $t_0 = 13.799 \pm 0.021\ Gyr$ in Table 4.1 (third column). *

> *This response supports the validity of the MR-model and the choice made of primary data.*

Remark 4.5 According to the ΛCDM-model[8] the age of the Universe is 13.73 ± 0.12 billion years, with (i) a Hubble constant $H_0 = 70.1 \pm 1.3\, kms^{-1} Mpc^{-1}$, (ii) 4. 6 % of ordinary baryonic matter; (iii) 23 % of dark matter of unknown nature; (iv) 72 % of dark energy favoring accelerating expansion; (v) less than 1 % of neutrinos. *

4.9 Analytical Profile of the MR-Model

Following the method of integration of a Weierstrass equation, we have constructed the numerical profile of the MR-Universe and obtained other significant numerical data. But what we really need in order to continue our analysis is the knowledge of an analytical profile, that is, to express the scale factor $a(t)$ by means of known elementary functions.

Searching for an exact solution of $a(t, t_0)$ of the Weierstrass equation (3.56) is not only a difficult problem, it is also unnecessary because it would lead to non-elementary transcendent functions that can be treated only by approximate representations. One might as well look for non-exact solutions involving elementary functions but capable of representing exact solutions with sufficient accuracy.

This idea can be realized by considering functions of type[9]

$$\boxed{a(t, t_0) = \alpha \sqrt[3]{\cosh(\beta\, t) - 1} = \alpha \sqrt[3]{\frac{1}{2}\, \frac{(e^{\beta\, t} - 1)^2}{e^{\beta\, t}}}} \tag{4.31}$$

where α and β are positive constants. Note that α must be dimensionless and $\text{Dim}\,(\beta) = \mathsf{T}^{-1}$.

---

[8] See the data provided by the *Wilkinson Microwave Anisotropy Probe project* (WMAP) [21].

[9] The two expressions in this profile are equivalent because $\cosh(x) - 1 = \frac{1}{2}\,(e^x + e^{-x} - 2) = \frac{1}{2}\, e^{-x}\left[e^{2x} + 1 - 2\, e^x\right] = \frac{1}{2}\, \frac{(e^x - 1)^2}{e^x}$..

Theorem 4.2 *Given the values of H_0 and t_0, the constants α and β are uniquely determined by imposing the two conditions*

(i) $a(t_0, t_0) = 1$ *(normalization condition),*

(ii) $H(t_0) = H_0$.

Proof Part 1. The Hubble factor of the profile (4.31) is

$$\boxed{H(t) \overset{\text{def}}{=} \frac{\dot{a}}{a} = \tfrac{1}{3}\,\beta\,\frac{e^{\beta t}+1}{e^{\beta t}-1}} \tag{4.32}$$

Indeed, setting for simplicity $X \overset{\text{def}}{=} e^{\beta t}$, we have successively

$$\begin{aligned} H(t) &= \frac{d\log a}{dt} = \tfrac{1}{3}\,\frac{d}{dt}\log\frac{(e^{\beta t}-1)^2}{e^{\beta t}} = \tfrac{1}{3}\,\frac{d}{dt}\log\frac{(X-1)^2}{X} \\ &= \tfrac{1}{3}\,\frac{d}{dt}\big[2\,\log(X-1) - \log X\big] = \tfrac{1}{3}\,(2\,\frac{\dot{X}}{X-1} - \frac{\dot{X}}{X}) \\ &= \tfrac{1}{3}\,\frac{2\,X - X + 1}{(X-1)\,X}\,\dot{X} = \tfrac{1}{3}\,\frac{X+1}{X-1}\,\frac{\dot{X}}{X} = \tfrac{1}{3}\,\frac{X+1}{X-1}\,\beta \quad \Longrightarrow (4.32). \end{aligned}$$

Part 2. Imposing condition (ii) $H(t_0) = H_0$ we find.

$$\boxed{\beta\,\frac{e^{\beta t_0}+1}{e^{\beta t_0}-1} = 3\,H_0} \tag{4.33}$$

This equation can be solved with respect to β because the function $H(t)$ (4.32) is increasing.

Part 3. Imposing condition (i) on (4.31) we find

$$\boxed{\alpha = \frac{1}{\sqrt[3]{\cosh(\beta\,t_0) - 1}} = \sqrt[3]{\frac{2\,e^{\beta t_0}}{(e^{\beta t_0}-1)^2}}} \tag{4.34}$$

Therefore, α is uniquely determined by β and t_0. □

Remark 4.6 From now on we will carry on the numerical analysis with the estimate (4.11) $\widehat{H}_0$ of the Hubble factor,

$$\boxed{H_0 = \widehat{H}_0 = 70.00\ km\ s^{-1} Mpc^{-1} \simeq 0.0715408\ Gyr^{-1}} \tag{4.35}$$

and consequently with the first estimate (4.24) of t_0:

$$\boxed{t_0 \simeq 13.3611603\ \ Gyr}\ * \tag{4.36}$$

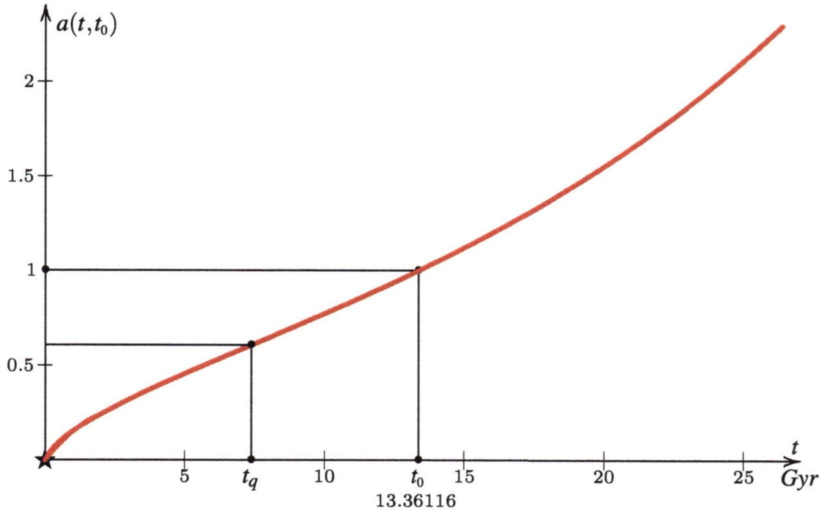

Fig. 4.6 Profile of the Universe in the MR-model

Theorem 4.3 *From the data* (4.35) *and* (4.36) *for the constants* α *and* β *we get the values*

$$\boxed{\alpha \simeq 0.607247} \quad \boxed{\beta \simeq 0.178366\ Gyr^{-1}} \tag{4.37}$$

Proof We calculate β by solving equation (4.33). Then α is calculated by applying (4.34). □

We then obtain two equivalent numerical expressions of the profile (4.31):

$$\boxed{a(t, t_0) = 0.607247 * \sqrt[3]{\cosh(0.178366 * t) - 1}} \tag{4.38}$$

$$\boxed{a(t, t_0) = 0.607247 * \sqrt[3]{\frac{1}{2} * \frac{[\exp(0.178366 * t) - 1]^2}{\exp(0.178366 * t)}}} \tag{4.39}$$

This analytical profile is plotted in Fig. 4.6 along with the pointwise numerical profile of Fig. 4.4: we see that ***they are virtually indistinguishable***.

The function (4.38) *is not an exact solution of the dynamical equation* (4.19) *but is a faithful representative of it that we will refer in the continuation of our analysis.*

4.10 Changing the Reference Time

If we like a profile with a reference time $t_\sharp$ different from today's t_0 we can apply the rules (1.25),

$$a(t, t_\sharp) = a(t, t_0)\, a(t_0, t_\sharp) = \frac{a(t, t_0)}{a(t_\sharp, t_0)},$$

so that from (4.31) we get

$$\boxed{a(t, t_\sharp) = \left[\frac{\cosh(\beta\, t) - 1}{\cosh(\beta\, t_\sharp) - 1}\right]^{\frac{1}{3}}} \tag{4.40}$$

In Fig. 4.7 we can see, for example, the two profiles related to $t_\sharp = 11$ and $t_\sharp = 17$. It is interesting to observe that the null acceleration time t_q (correponding to the inflection points) is invariant.

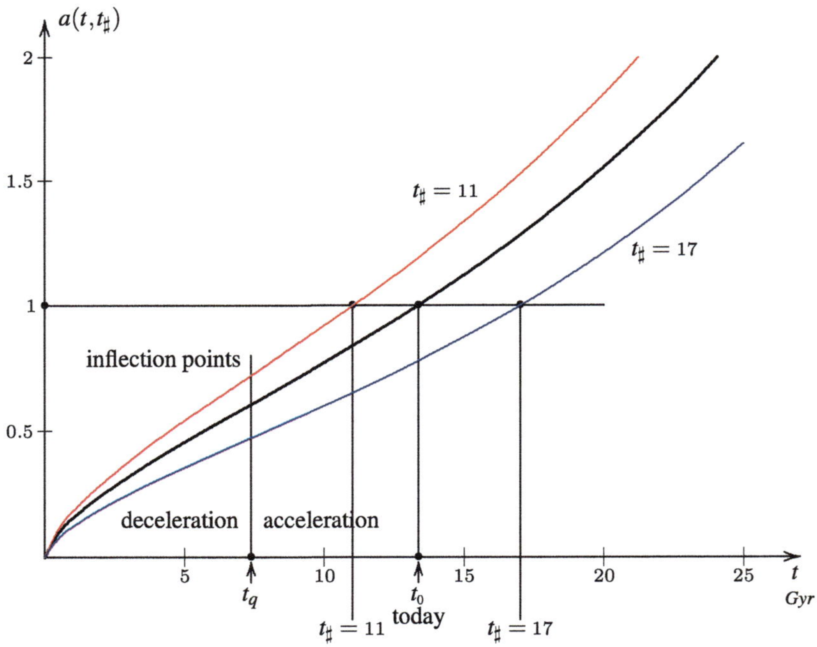

Fig. 4.7 Profiles with reference times $t_\sharp$ different from t_0

4.11 Time Evolution of the Hubble Factor

The function $H(t)$ (4.32) admits the equivalent analytical representation

$$\boxed{H(t) = \tfrac{1}{3}\,\beta\,\frac{\sinh(\beta\,t)}{\cosh(\beta\,t) - 1}} \tag{4.41}$$

where the constant α is non longer involved. Also (Fig. 4.8):

$$\begin{cases} \lim\limits_{t\to 0} H(t) = \tfrac{1}{3}\,\beta\,\lim\limits_{t\to 0}\dfrac{\cosh(\beta\,t)}{\sinh(\beta\,t)} = +\infty. \\ H_\infty = \lim\limits_{t\to+\infty} H(t) = \tfrac{1}{3}\,\beta\,\lim\limits_{t\to+\infty}\dfrac{\cosh(\beta\,t)}{\sinh(\beta\,t)} = \tfrac{1}{3}\,\beta = \underline{0.0594556}. \end{cases}$$

The numerical expression of $H(t)$ (4.32) is then

$$\boxed{H(t) = 0.0594556 * \frac{e^{0.178366*t} + 1}{e^{0.178366*t} - 1}} \tag{4.42}$$

Remark 4.7 Equation (4.32) is solvable with respect to $X = e^{\beta\,t}$:

$$\left[\begin{array}{l} H = \tfrac{1}{3}\,\beta\,\dfrac{X+1}{X-1} \Longrightarrow 3\,H\,(X-1) = \beta\,(X+1) \\ \Longrightarrow (3\,H - \beta)\,X = \beta + 3\,H \end{array}\right. \Longrightarrow X = \frac{3\,H + \beta}{3\,H - \beta}.$$

Since $X = e^{\beta\,t}$, it follows that

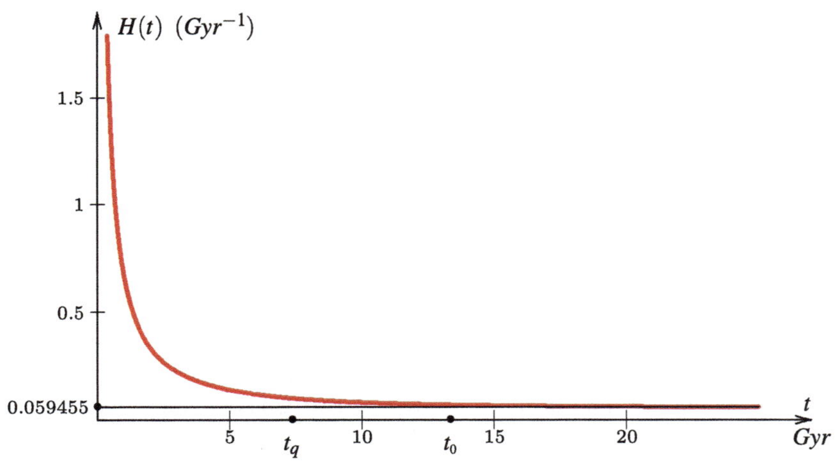

Fig. 4.8 Evolution of the Hubble factor $H(t)$

$$t = \frac{1}{\beta} \log \frac{3\,H(t) + \beta}{3\,H(t) - \beta} \tag{4.43}$$

The evaluation for $t = t_0$ gives a new expression of the age of the Universe

$$t_0 = \frac{1}{\beta} \log \frac{3\,H_0 + \beta}{3\,H_0 - \beta} \tag{4.44}$$

in terms of the constants H_0 and β. Plugging in the data (4.11) $H_0 = \widehat{H}_0 = 0.07154\;Gyr^{-1}$ and (4.37) $\beta = 0.178366\;Gyr^{-1}$ gives the estimate

$$t_0 = 13.36116\;Gyr \tag{4.45}$$

in perfect agreement with that previously found by means of the integral

$$t_0 = \int_0^1 \frac{dx}{\sqrt{W(x)}}$$

This result attests the reliability of the MR-model. *

4.12 Super-Luminal Recession Speed

Combining Hubble law (1.34) $\dot{d}_{AB}(t) = H(t)\,d_{AB}(t)$ with equation $d_{AB}(t) = a(t, t_0)\,d_{AB}(t_0)$ derived from (1.30) by posing $t_\sharp = t_0$, we get the recession speed of two galaxies

$$\dot{d}_{AB}(t) = \dot{a}(t, t_0)\,d_{AB}(t_0) \tag{4.46}$$

in terms of their present time distance $d_{AB}(t_0)$ and the growth rate of the scale factor $\dot{a}(t, t_0)$.

In the MR-model this growth rate is given by

$$\dot{a}(t, t_0) = \tfrac{1}{3}\,\alpha\,\beta\,\frac{\sinh(\beta\,t)}{\left[\cosh(\beta\,t) - 1\right]^{\frac{2}{3}}} \tag{4.47}$$

It is obtained by deriving equation (4.31) with respect to t. It is plotted in Fig. 4.9.

Given the present time distance $d_{AB}(t_0)$ of two galaxies, the recession speed $\dot{d}_{AB}(t)$, which is unbounded around $t = 0$, decreases rapidly to a minimum value 0.06253447 at time $t_q \simeq 7.37949$, after which it begins to increase slowly. The time

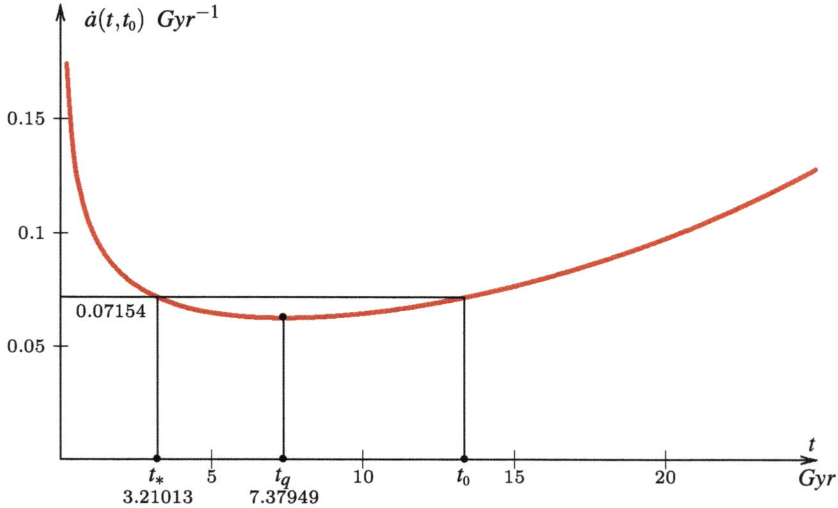

Fig. 4.9 The growth rate $\dot{a}(t, t_0)$

of minimum t_q does not depend on the distance $d_{AB}(t_0)$ and is determined by the equation $\ddot{a}(t_q, t_0) = 0$. It is therefore the time of the beginning of the accelerated expansion.

The recession speed $\dot{d}_{AB}(t)$ of two galaxies can be **super-luminal**, i.e. it could exceed the speed of light:

$$\dot{d}_{AB}(t) \geq c. \tag{4.48}$$

This circumstance does not contradict the canons of relativity because a recession speed is not the speed of a particle with respect to a reference frame, but is due to the expansion-contraction of the Universe.

In the units *Gyr* and *Glyr* that we have used so far for times and lengths, the numerical value of the light speed is $c = 1$ (Sect. 4.4). Thus, due to (4.46), the **super-luminal condition** (4.48) is expressed by the inequality

$$\boxed{d_{AB}(t_0) \geq \frac{c}{\dot{a}(t, t_0)}, \quad c = 1} \tag{4.49}$$

Consequently,

Theorem 4.4 *Two cosmic bodies with current distance $d_{AB}(t_0)$ have a super-luminal recession speed in the time-interval where the inequality* (4.49) *is satisfied.*

The super-luminal condition (4.49) shows that the function

$$\boxed{\mathsf{L}(t, t_0) \stackrel{\text{def}}{=} \frac{c}{\dot{a}(t, t_0)}, \quad c = 1} \tag{4.50}$$

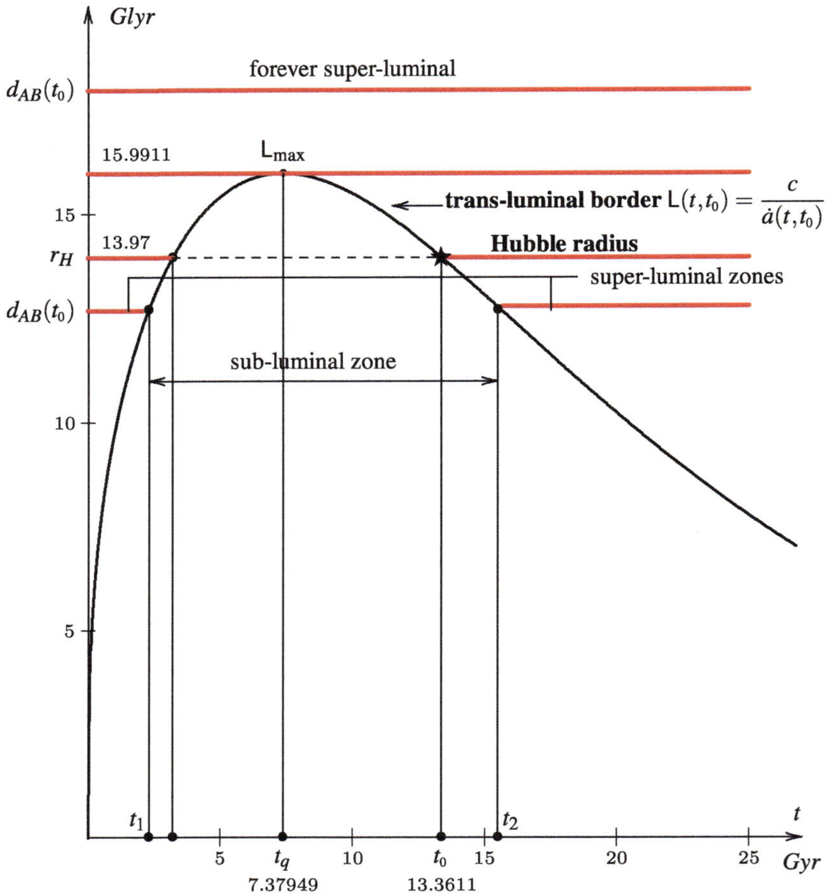

Fig. 4.10 Trans-luminal border $L(t, t_0)$ and recession speed

plays the role of **trans-luminal border** marking the transition from sub-luminal to super-luminal states. The meaning of Theorem 4.4 is explained by Fig. 4.10.
1. The transluminal border attains its maximum value at the time t_q, when the acceleration of the scale factor is zero. In fact, from the definition (4.50) it follows that

$$\frac{dL(t, t_0)}{dt} = \frac{d}{dt}\frac{c}{\dot{a}(t, t_0)} = -c\,\frac{\ddot{a}(t, t_0)}{\dot{a}^2(t, t_0)}$$

and this derivative is zero when $\ddot{a}(t, t_0) = 0$. The maximum value turns out to be

$$\boxed{L_{max} \simeq 15.9911\ Glyr} \tag{4.51}$$

2. Consider two cosmic bodies A and B whose current distance is $d_{AB}(t_0)$ and draw a horizontal line corresponding to this value.
3. If $d_{AB}(t_0) < \mathsf{L}_{\max}$ then this line crosses $\mathsf{L}(t, t_0)$ in two points which correspond to two times $t_1 < t_2$. Consequently, the recession speed is sub-luminal in the temporal interval (t_1, t_2). and super-luminal outside this interval.
4. If $d_{AB}(t_0) > \mathsf{L}_{\max}$ then the recession speed of the two bodies is always super-luminal.
5. Close to the big-bang the recession speed of whatever couple of galaxies is super-luminal.
6. The astronomers have introduced the **Hubble radius** (or **Hubble length**) defined by

$$\boxed{r_H \overset{\text{def}}{=} \frac{c}{H_0}} \tag{4.52}$$

In our context this length is equal to the current value of the transluminal function, i.e.

$$\boxed{r_H = \mathsf{L}(t_0, t_0)} \tag{4.53}$$

This follows by applying formula (1.35) written for $t_\sharp = t_0$, i.e. $H(t_0) = \dot{a}(t_0, t_0)$:

$$\mathsf{L}(t_0, t_0) = \frac{c}{\dot{a}(t_0, t_0)} = \frac{c}{H_0} = r_H.$$

Furthermore, observing the location of the point ★ in Fig. 4.10 we can state that

Theorem 4.5 *If the today distance $d_{AB}(t_0)$ of two galaxies A and B is equal to the Hubble radius r_H then their recession speed starts today to be super-luminal and will remain so forever.*

4.13 Models with Null Cosmological Constant

For $\Lambda = 0$ the Weierstrass equation (3.43) becomes

$$\dot{a}^2 = H_0^2 \left[\Omega_m\, a^{-1} + \Omega_r\, a^{-2} \right] - c^2\, K_0. \tag{4.54}$$

Theorem 4.6 *In the $\Lambda = 0$ models, the space curvature cannot be zero.*

Proof With $\Lambda = 0$ Eq. (3.54) reduces to

$$\Omega_m(t_0) + \Omega_r(t_0) = 1, \tag{4.55}$$

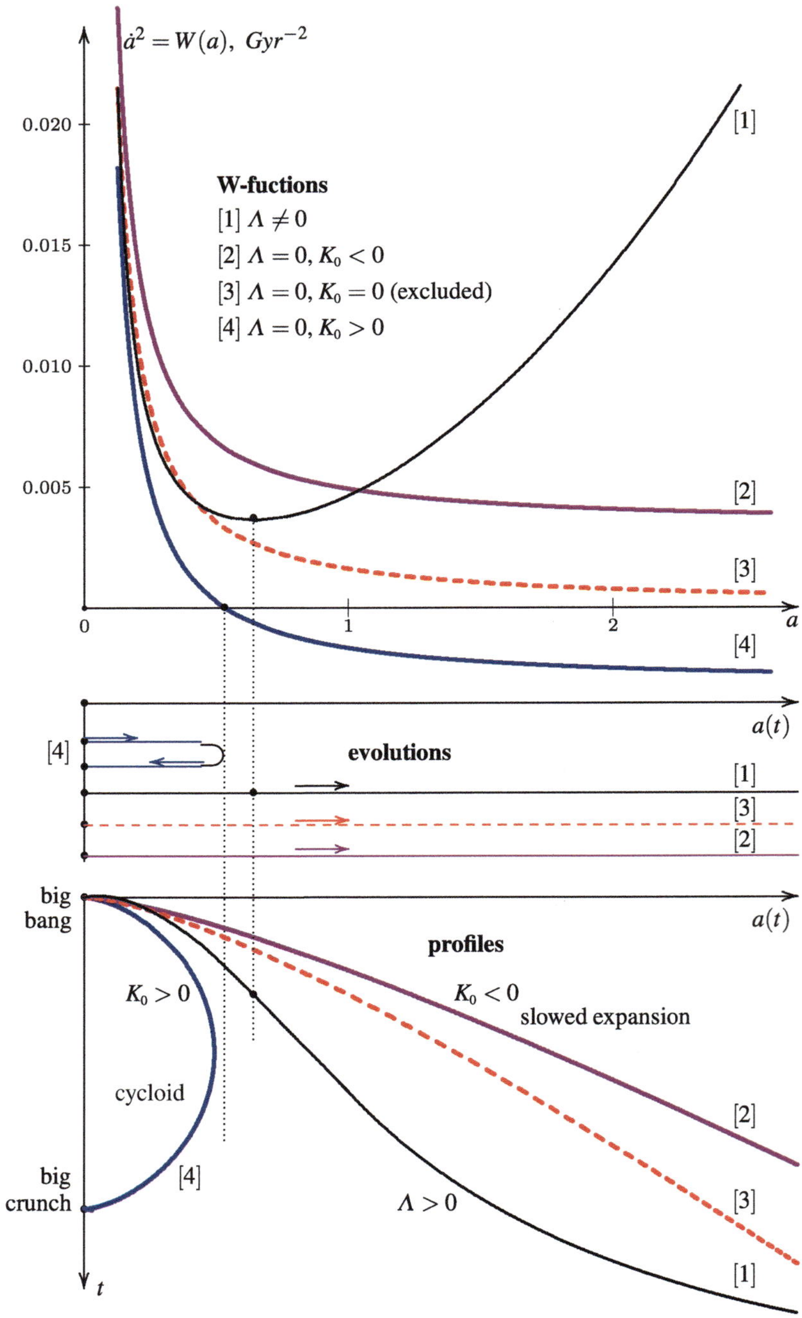

Fig. 4.11 Comparison between profiles with $\Lambda = 0$ and $\Lambda \neq 0$

and with the same arguments as in the proof of Theorem 3.7 we arrive at an absurdity. □

Figure 4.11 compares the two functions (4.54) corresponding to the two signs of K_0 and the Weierstrass function (4.19) of the MR-model.

For $K_0 < 0$, it is always $W(a) > 0$: no stopping point. The Universe has a slowly decelerating expansion from the beginning until $+\infty$.

For $K_0 > 0$, $W(a)$ has a simple zero. The Universe has a cycloidal-type evolution with a big-bang and a big-crunch in finite time.

We observe that in the case $K_0 > 0$ there is a discontinuous change of profiles in the transition from the $\Lambda = 0$ model to the $\Lambda > 0$ model. *This fact gives the model* $\Lambda = 0$ *an unacceptable character of instability.*

For $K_0 > 0$ the profile of the Universe is similar to a cycloid. In fact, it is a true cycloid in the case where the radiation density is neglected. In fact, if we place $\Omega_r = 0$ in the equation (4.54)

$$\dot{a}^2 = H_0^2\, \Omega_m\, a^{-1} - c^2\, K_0$$

it can be verified that the solutions are expressed by the parametric equations

$$\begin{cases} c\,t = \dfrac{\mathsf{B}}{\sqrt{K_0}}\,(\theta - \sin\theta), \\ a = \mathsf{B}\,(1 - \cos\theta), \end{cases} \qquad \mathsf{B} \overset{\text{def}}{=} \frac{1}{2}\,\frac{H_0^2}{c^2}\,\frac{\Omega_m}{K_0},$$

it can be verified that the solutions are expressed by the parametric equations

$$\begin{cases} c\,t = \mathsf{B}\,r_0\,(\theta - \sin\theta), \\ r = \mathsf{B}\,r_0\,(1 - \cos\theta). \end{cases} \tag{4.56}$$

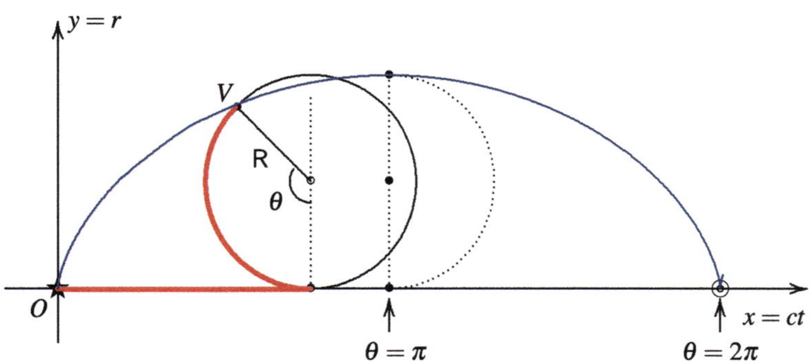

Fig. 4.12 Evolution of the cosmic radius r for $K_0 > 0$ and $\Omega_r = 0$

In the plane $(x, y) = (ct, r)$ these equations represent the cycloid described by a point V on the edge of a circle of radius $\mathsf{R} = \mathsf{B}\, r_0$ rolling on the x axis, as shown in Fig. 4.4. The parameter θ is the angle of rotation of the circle, with $\theta = 0$ for $t = 0$ (Fig. 4.12).

Chapter 5
Signal Transmission and Visibility

In Sect. 1.12 we identified the reference space with a *fabulous instrument*, called **cosmic monitor**, that provides a virtual view of the Universe where galaxies are fixed points and wandering particles are visible as curves parameterized by cosmic time t. We will make an extensive use of this tool in this and the next chapter.

5.1 Photon Transmission

Photons are very strange particles: they are never at rest and have the same speed c with respect to any reference space. This is in agreement with the theory of propagation of electromagnetic waves: by virtue of Theorem 2.9, in space-time the histories of photons are identified with electromagnetic rays, of visible or non-visible frequency. Therefore, photons (in a broad sense) are the vehicle by which signals are transmitted, and therefore we will consider the terms **photon** and **signal** as synonyms.

The current view of the primordial Universe places at about 377000 years after the big-bang the transition from an opaque cosmos to a transparent one. In this epoch, called **recombination**, the first neutral atoms formed and reached their minimum energy state generating photons (**photon decouplig**) that even today can be intercepted as **cosmic background radiation**. But at that time stars have not yet formed and there are no light sources. The first stars and galaxies formed around 400-700 million years after the big-bang.

To work on this topic we associate with each galaxy A an **initial emission time** t_{bA} at which A begins to emit photons. All these times belong to a time interval called **reionization epoch** whose beginning is, see (4.29),

$$\boxed{t_{\mathrm{re}} \simeq 0.54409\ Gyr} \tag{5.1}$$

S. Benenti, *Mathematical Foundations and Numerical Analysis of the Dynamics of an Isotropic Universe*, https://doi.org/10.1007/978-3-031-64033-9_5

The relativistic bridge-postulate states that photons are characterized by equation

$$\boxed{a(t, t_\sharp)\,\frac{ds_\sharp}{dt} = c}$$

where $ds_\sharp$ is the arc-element of the reference space. This equation implies that
(i) If $\lim_{t\to t_\alpha} a(t) = 0$ (t_α beginning of Universe) then

$$\lim_{t\to t_\alpha} \frac{ds_\sharp}{dt} = +\infty.$$

This means that[1]

Theorem 5.1 *If* $\lim_{t\to t_\alpha} a(t) = 0$ *then on the cosmic monitor the speed* $\dot{s}_\sharp$ *of a photon tends to* $+\infty$ *when t tends to the beginning of Universe.*

Note that $\dot{s}_\sharp$ is not the peculiar speed.

(ii) If $\lim_{t\to t_\omega} a(t) = +\infty$ (t_ω end of the Universe) then

$$\lim_{t\to t_\omega} \frac{ds_\sharp}{dt} = 0.$$

This means that *on the cosmic monitor, the speed of a photon tends to zero when t tends to the end of the Universe*. In other words, its trajectory tends toward a fixed point, that is, a galaxy. In turn, this is equivalent to saying

Theorem 5.2 *If* $\lim_{t\to t_\omega} a(t) = +\infty$ *then, in approaching the end of the Universe, the history of a photon tends to touch the history of a galaxy.*

Remark 5.1 Both assumptions of these theorems are satisfied in the MR-model.

5.2 Emission-Reception Relationship

This chapter will deal with two cosmic bodies (galaxies) A and B whose histories are time-like geodesics belonging to the cosmic fluid congruence:

(i) body A is capable of emitting photons whose histories are null geodesics of space-time;

(ii) body B is equipped with instruments capable of intercepting photons from the cosmos.

Suppose that a photon P is emitted from A at a certain **emission time** t_{eA} and that its history crosses the history of B at a **reception time** t_{rB}, as shown in Fig. 5.1.

[1] See also Theorem 2.6 and Remark 2.2.

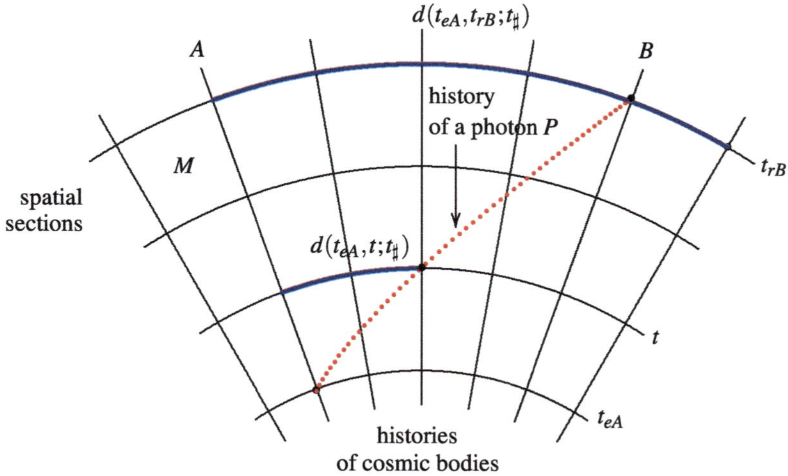

Fig. 5.1 History of a photon P emitted by A and received from B

Fig. 5.2 Progression of photon P observed on the cosmic monitor

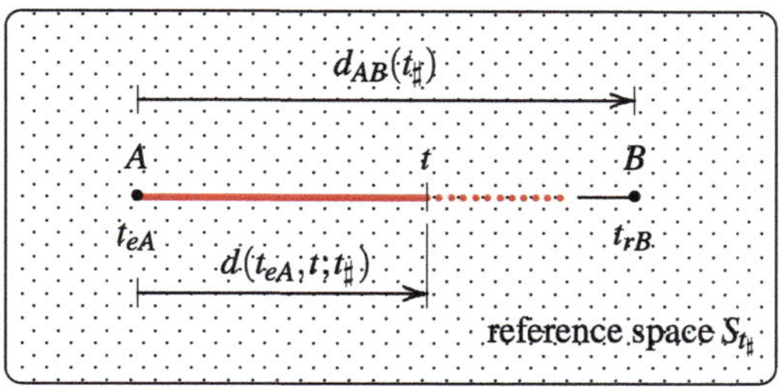

Theorem 5.3 *A photon emitted from A at time t_{eA} can reach B at a time t_{rB} if and only if*

$$\boxed{c\int_{t_{eA}}^{t_{rB}} \frac{dt}{a(t, t_\sharp)} = d_{AB}(t_\sharp)} \tag{5.2}$$

where $d_{AB}(t_\sharp)$ is the distance of A from B measured in the reference space $S_{t_\sharp}$.

We call Eq. (5.2) **emission-reception relationship**.

Proof. The progression of P on the cosmic monitor is shown in Fig. 5.2.

Since the trace of P is a geodesic, the trajectory of P is represented by a line segment with endpoints A to B. Its length is $d_{AB}(t_\sharp)$. Let us denote by

$$d(t_{eA}, t; t_\sharp)$$

the distance traveled by P at time t. From Eq. (2.7) we infer that the point P moves with velocity

$$\frac{ds_\sharp}{dt} = \frac{c}{a(t, t_\sharp)}. \tag{5.3}$$

Then the distance traveled by P at any time $t > t_{eA}$ is given by

$$\boxed{d(t_{eA}, t; t_\sharp) = c \int_{t_{eA}}^{t} \frac{dt}{a(t, t_\sharp)}} \tag{5.4}$$

For $t = t_{rB}$ we find $d(t_{eA}, t_{rB}; t_\sharp) = d_{AB}(t_\sharp)$. ∎

Remark 5.2 The integral

$$\eta(t; t_\sharp) = \int_0^t \frac{dt}{a(t, t_\sharp)} \tag{5.5}$$

with $t_\sharp = t_0$ Is called **conformal time** by cosmologists.

Remark 5.3 The emission-reception relationship (5.2) has a remarkable geometrical interpretation illustrated in Fig. 5.3.

The shaded area delimited by the graph of $c/a(t, t_\sharp)$ over the emission-reception interval $[t_{eA}, t_{rB}]$ is equal to the integral (5.2) and thus equal to the reference distance $d_{AB}(t_\sharp)$. Note that, from the dimensional point of view, this 'area' is in fact a time × velocity, that is, it has the dimension of a length.

Theorem 5.4 *The emission-reception relationship (5.2) does not depend on the choice of reference time* $t_\sharp$.

Proof. By virtue of (1.27), $a(t, t_\sharp) = a(t, t_\flat)\, a(t_\flat, t_\sharp)$, we have

$$d(t_{eA}, t; t_\sharp) = c \int_{t_{eA}}^{t} \frac{dt}{a(t, t_\sharp)} = \frac{c}{a(t_\flat, t_\sharp)} \int_{t_{eA}}^{t} \frac{dt}{a(t, t_\flat)} = a(t_\sharp, t_\flat)\, d(t_{eA}, t; t_\flat).$$

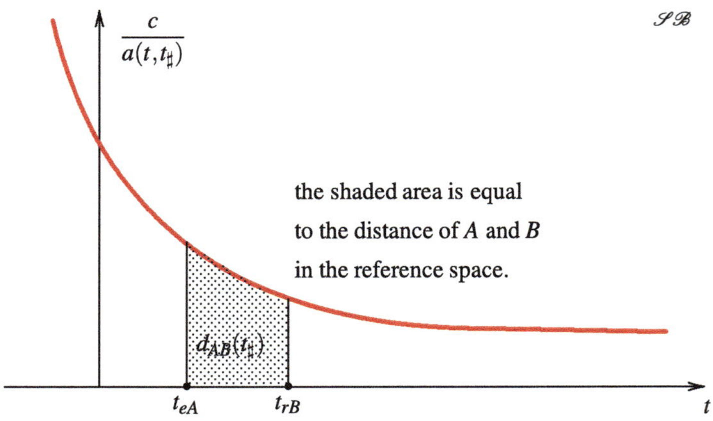

Fig. 5.3 Geometrical interpretation of Eq. (5.2)

Subsisting the relation (1.30) $d_{AB}(t_\sharp) = a(t_\sharp, t_\flat)\, d_{AB}(t_\flat)$ between the distances, it follows that

$$d(t_{eA}, t; t_\flat) = d_{AB}(t_\flat)$$

for any t, specifically for $t = t_{rB}$. ■

Because of this independence we can take the present time t_0 not only as the reference time but also as the receception time t_{rB}. Then the emission-reception relationship (5.2) becomes.

$$\boxed{c \int_{t_{eA}}^{t_0} \frac{dt}{a(t, t_0)} = d_{AB}(t_0)} \tag{5.6}$$

This formula gives today's distance of a galaxy A from B when the observer in B knows the emission time t_{eA} of the photon and also the profile $a(t, t_0)$ of the Universe.

For the MR-model, the profile is given by the (4.38)

$$a(t, t_0) = \alpha \sqrt[3]{\cosh(\beta\, t) - 1} \qquad \begin{cases} \alpha \simeq 0.607247 \\ \beta \simeq 0.178366\, Gyr^{-1}. \end{cases}$$

The function under integration in the (5.6) is plotted in Fig. 5.4. In accordance with the geometrical interpretation shown in Fig. 5.3, the shaded area multiplied by c gives the current distance $d_{AB}(t_0)$.

Remark 5.4 If in Eq. (5.6) we place as the lower margin of integration the reionization time $t_{re} = 0.54409$, that is, the time at which cosmic bodies begin to emit light, then *we get the measure of the maximum distance of celestial bodies observable today*:

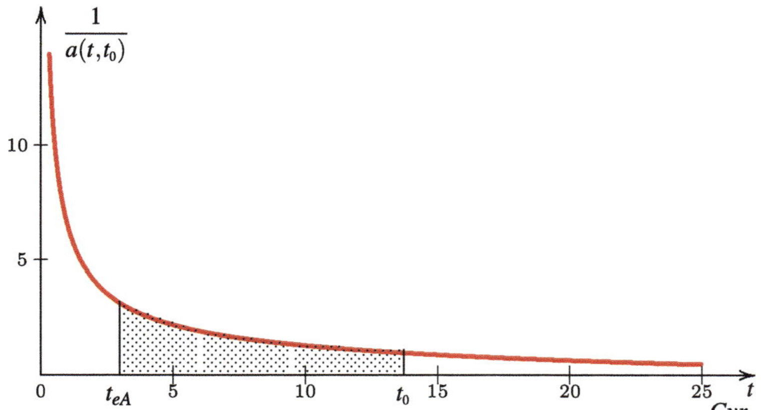

Fig. 5.4 The function $1/a(t, t_0)$ in the MR-model

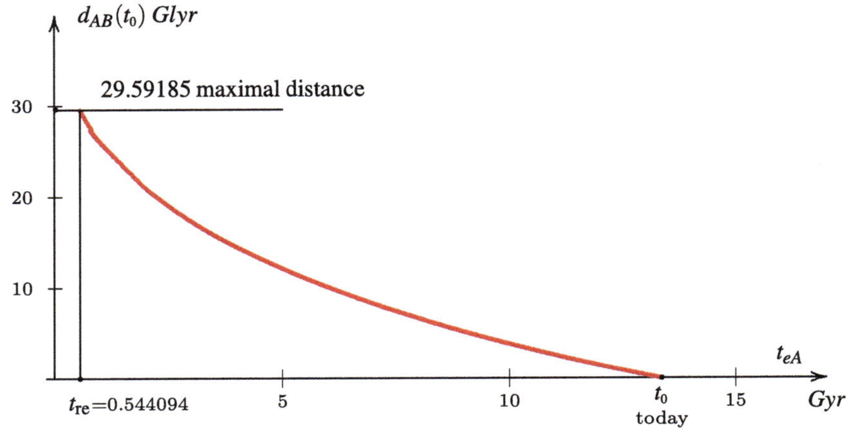

Fig. 5.5 Emission time t_{eA} vs. today's distance $d_{AB}(t_0)$

Table 5.1 $t_{eA} \mapsto d_{AB}(t_0)$.

t_{eA}	$d_{AB}(t_0)$	t_{eA}	$d_{AB}(t_0)$
t_{re}	29.59185	6	10.09304
0.6	29.06055	7	8.27914
0.7	28.18729	8	6.65050
0.8	27.39381	9	5.17287
1	25.98672	10	3.82179
2	20.89091	11	2.57909
3	17.32891	12	1.43079
4	14.50813	13	0.36588
5	12.14273	t_0	0

$$\boxed{\max d_{AB}(t_0) = c \int_{t_{re}}^{t_0} \frac{dt}{a(t, t_0)} \simeq 29.59185\ Glyr} \quad \bullet \tag{5.7}$$

Remark 5.5 Equation (5.6) can be used to solve the inverse problem: *knowing the distance $d_{AB}(t_0)$ determine the emission time t_{eA}* (see Fig. 5.5 and Table 5.1).

5.3 Visible and Unvisible Universe

We say that A is **visible** from B at date t if on that date B intercepts photons from A. We assume that *every body A has its own* **initial emission time** t_{bA}, *time at which it begins to emit photons* which we call **primordial photons**.

Fig. 5.6 Progression in S_{t_0} of a primordial photon P_A

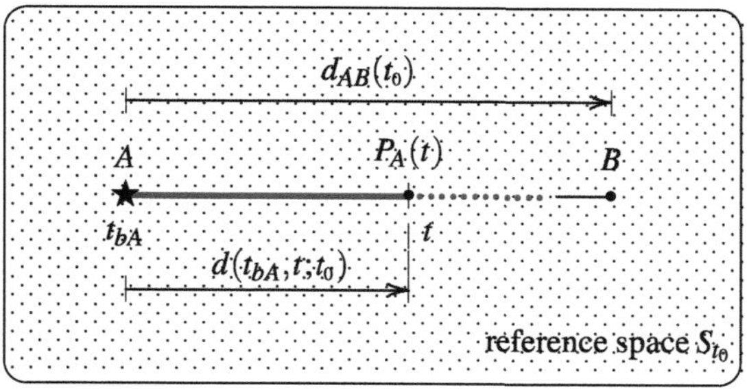

Fig. 5.7 Theorem 5.5

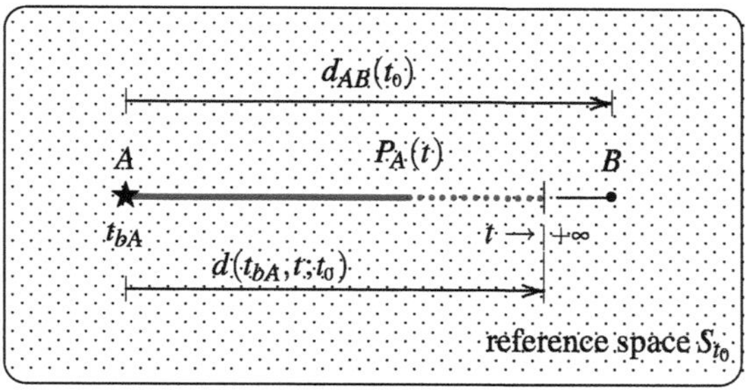

Let us observe the progression of a primordial photon P_A in today's reference space S_{t_0}.[2]

Note that (Fig. 5.6):

(i) In the reference space, A and B are fixed points, whose distance is $d_{AB}(t_0)$.

(ii) P_A travels on a line from A toward B. This line is the projection on S_{t_0} of the history of P_A in space-time, which is a light-like geodesic. The distance measured in S_{t_0} and traveled in any time $t > t_{bA}$ is given by (see Eq. (5.4))

$$\boxed{d(t_{bA}, t; t_0) = c \int_{t_{bA}}^{t} \frac{dt}{a(t, t_0)}} \tag{5.8}$$

This is an increasing function of t that we call **progression of a primordial photon**.

Theorem 5.5 *If the function $d(t_{bA}, t; t_0)$ is bounded and if*

$$d_{AB}(t_0) > \lim_{t \to +\infty} d(t_{bA}, t; t_0)$$

then P_A can never reach B.

Proof. Figure 5.7 is self-explanatory. ∎

Theorem 5.6 *If $d_{AB}(t_0) = d(t_{bA}, t_0; t_0)$ then A begins today to be visible from B.*

[2] Put $t_\sharp = t_0$ and $t_{eA} = t_{bA}$ in Fig. 5.2.

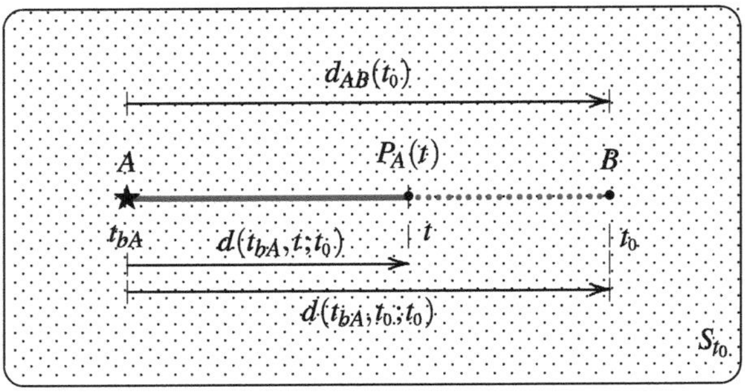

Fig. 5.8 Theorem 5.6

Proof Figure 5.8 is self-explanatory. ■

Recall that the **reionization time** t_{re} is the minimum of all the times t_{bA} of primordial photon emission. Therefore, the distance

$$\boxed{r_{vis}(t) \stackrel{\text{def}}{=} d(t_{re}, t; t_0) = c \int_{t_{re}}^{t} \frac{dt}{a(t, t_0)}} \tag{5.9}$$

obtained from equation (5.8) by replacing t_{bA} with the reionization time t_{re} is the *maximum distance of a galaxy A visible from B at time t*. We call it **visibility radius of the Universe at time** t. The use of the term 'radius' will be clarified shortly.

Note that this **visibility relation** is symmetrical with respect to A and B.

The numerical representation of the radius of visibility in the MR-model (whose profile is given in (4.38)) is[3]

$$\boxed{r_{vis}(t) = \frac{c}{0.607247} \int_{t_{re}}^{t} \frac{dt}{\sqrt[3]{\cosh(0.178366 * t) - 1}}} \tag{5.10}$$

The graph is plotted in Fig. 5.9 where two significant numbers are highlighted:

• Current value of the visibility radius

$$\boxed{r_{vis}(t_0) \simeq 29.59185\ Glyr} \tag{5.11}$$

This is the distance $d_{AB}(t_0)$ of a body A which today begins to be visible from B. We call this the **current visibility radius**.

•• The limit

$$\boxed{r_{inv} \stackrel{\text{def}}{=} \lim_{t \to +\infty} r_{vis}(t) \simeq 45.61382\ Glyr} \tag{5.12}$$

[3] In this formula we must put $c = 1$ for the result to be in $Glyr$ units (Sect. 4.4).

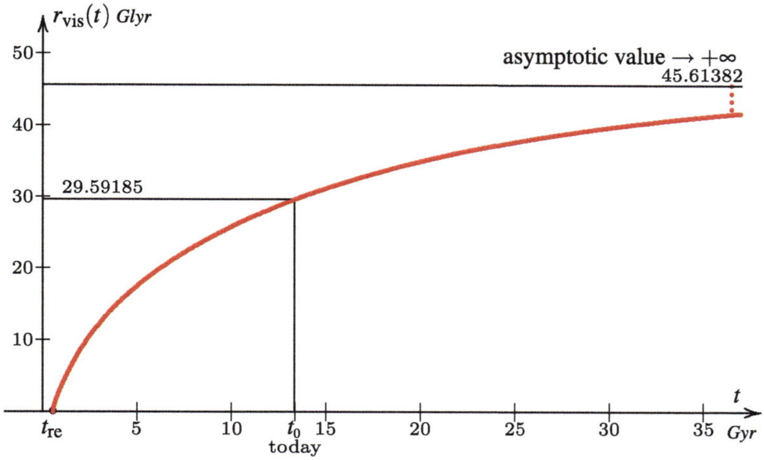

Fig. 5.9 Radius of the visible Universe $r_{\text{vis}}(t)$

Since $r_{\text{inv}} = \lim_{t\to+\infty} d(t_{\text{re}}, t; t_0)$ then, according to Theorem 5.5 applied to the case $t_{bA} = t_{\text{re}}$, if

$$d_{AB}(t_0) > r_{\text{inv}} \tag{5.13}$$

then no photon emitted by any galaxy A will reach B in the future. In other words, an observer in B will never see galaxies A whose current distance $d_{AB}(t_0)$ is greater than $r_{rminv} \simeq 45.61382$ *Glyr* (5.12). This is the **absolute invisibility radius of the Universe**, that is, the *radius of the Universe forever invisible to any observer B*.

The progressions $d(t_{bA}, t; t_0)$ of the primordial photons for some values of t_{bA} (2.5, 5, 7.5, 10) are plotted in Fig. 5.10. The number $d(t_{bA}, t_0; t_0)$ is the current distance $d_{AB}(t_0)$ of the galaxy A beginning today to be visible from B (Theorem 5.6).

Remark 5.6 The radius $r_{\text{vis}}(t)$ is called by cosmologists **particle horizon** or even **cosmological horizon**. Because of the expansion of the Universe, its present-day value $r_{\text{vis}}(t_0)$ is not simply the age of the Universe t_0 multiplied by the speed of light, as for the Hubble horizon (see note below) but rather the speed of light multiplied by the conformal time, as can be seen by combining (5.9) with (5.5). Today's *particle horizon* represents the extreme distance from which we can draw information about the Universe's past. •

Remark 5.7 The **Hubble radius** is also called **Hubble horizon**.

It establishes the boundary between particles that are always moving slower than the speed of light relative to an observer at a given time. •

Remark 5.8 Cosmologists call **event horizon** the minimum distance from which light emitted now can ever reach the observer in the future. Due to (5.13) this distance coincides with our radius of absolute invisibility r_{inv}.

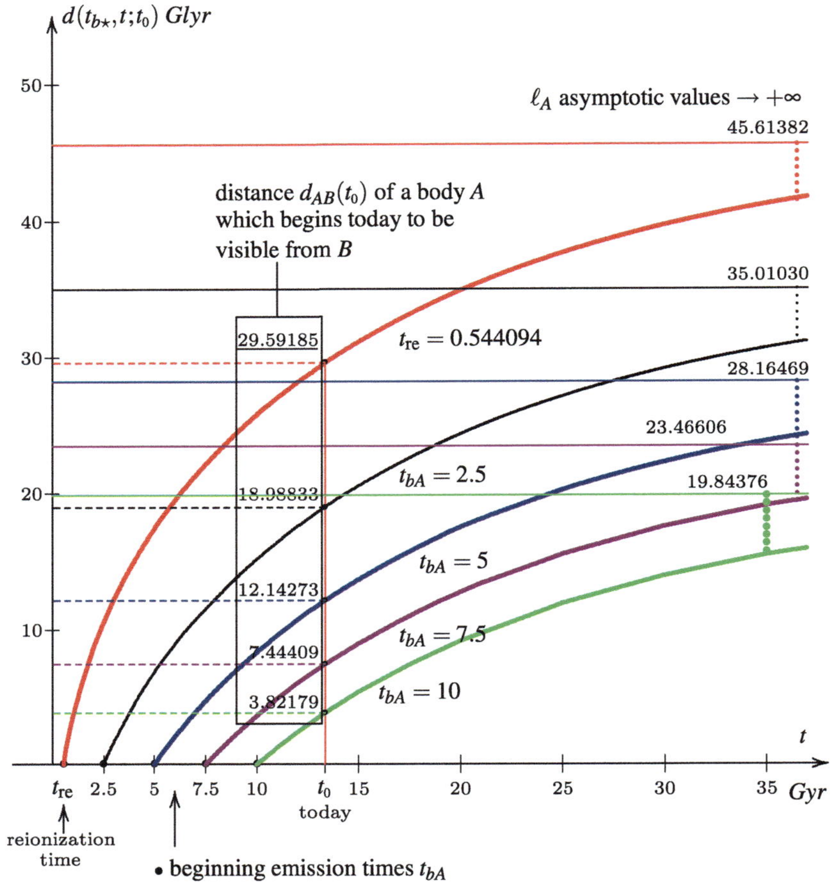

Fig. 5.10 Primordial photon progressions $d(t_{bA}, t; t_0)$

5.4 Present-Time Configuration of the Universe

Recall what has already been said in the Preface.

The large circle that appears in Figure 0.4, reprinted here from the end of the preface, represents the three-dimensional sphere $\mathbb{S}_3$ of radius $r_0 \simeq 1460.42$ *Glyr* (4.15) where the galaxies are currently distributed. Because of the extremely small curvature of this sphere, to an observer in any galaxy B, the Universe appears flat at least in a neighborhood of about 29.59 billion light-years, as shown in the upper part of the figure. This distance is the **radius of current visibility** $r_{vis}(t_0)$ (5.11) of the Universe. This radius is understood to be measured on the sphere $\mathbb{S}_3$ and on this determines a spherical cap of semi-amplitude $\psi_{vis}(t_0)$.

In turn, the **invisibility radius of the Universe** $r_{inv} \simeq 45.61$ *Glyr* (5.12), beyond which the Universe remains forever invisible to B, results in a spherical cap of half-amplitude ψ_{inv}.

Both angles $\psi_{vis}(t_0)$ and ψ_{inv} are not shown to scale with respect to the rest of the drawing. They are actually very small and would therefore be almost unnoticeable:

$$\begin{cases} \psi_{vis}(t_0) = \dfrac{r_{vis}(t_0)}{r_0} = \dfrac{29.59185}{1460.429942} \simeq 0.020262 \\ \psi_{inv} = \dfrac{r_{inv}}{r_0} = \dfrac{45.61382}{1460.429942} \simeq 0.031233 \end{cases} \tag{5.14}$$

In the course of the expansion of the Universe, the angle ψ_{inv} remains unchanged, while ψ_{vis} grows asymptotically to the limit ψ_{inv} for $t \to +\infty$ (Fig. 5.11).

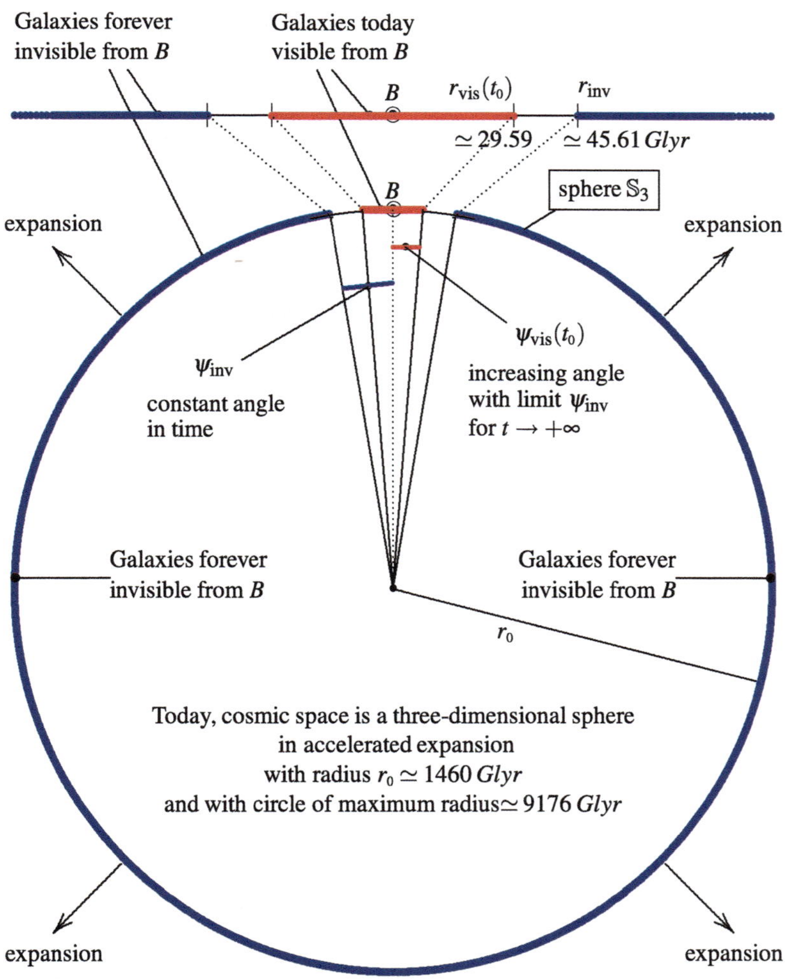

Fig. 5.11 Present time Universe configuration referred to an observer placed in a galaxy *B*

On a sphere $\mathbb{S}_3$ with radius r the volume of a spherical cap of semi-amplitude ψ is given by (see Eq. (7.9), Figure 7.3)

$$\boxed{V(r,\psi) = 2\pi\, r^3\,(\psi - \sin\psi\,\cos\psi)} \tag{5.15}$$

For $\psi = \pi$ we get the volume of the whole sphere $\mathbb{S}_3$

$$\boxed{V = 2\pi^2 r^3} \tag{5.16}$$

It follows that the today volume of the whole Universe is

$$\boxed{VU(t_0) = 2\pi^2 r_0^3 \simeq 61.485 * 10^9\; Glyr^3} \tag{5.17}$$

whereas that of the visible Universe is

$$\boxed{VU_{\rm vis}(t_0) = 2\pi\, r_0^3\,(\psi_{\rm vis} - \sin\psi_{\rm vis}\,\cos\psi_{\rm vis}) \simeq 108534.8\; Glyr^3} \tag{5.18}$$

5.5 Visibility and Super-Luminal Recession Speed

Figure 5.12 shows both graphs of the visibility radius $r_{\rm vis}(t)$ (Fig. 5.9) and the trans-luminal boundary $\mathsf{L}(t, t_0)$ (Fig. 4.10).

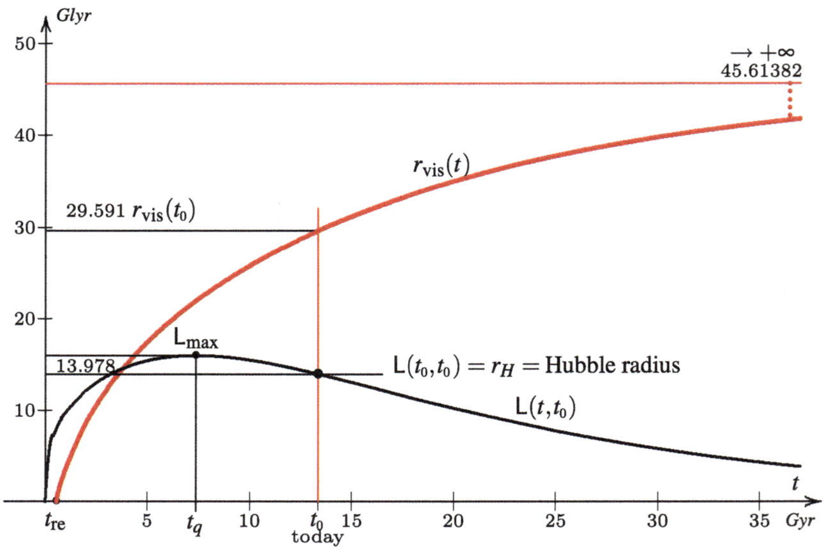

Fig. 5.12 Radius $r_{\rm vis}(t)$ of the visible Universe and trans-luminal boundary $\mathsf{L}(t, t_0)$ compared

We observe that the vertical line in t_0 intersects the graph of $\mathsf{L}(t, t_0)$ at the level $r_H = \mathsf{L}(t_0, t_0)$ corresponding to the Hubble radius, as already shown in Sect. 4.12. Recalling Remark 5.7 we conclude that

If the current distance of two galaxies A and B is greater than the Hubble radius

$$d_{AB}(t_0) > r_H$$

then A and B have super-luminal recession velocities even though they are mutually visible.

Interesting paradox.

Chapter 6
Cosmic Redshift

The cosmic redshift is a phenomenon, due to the expansion or contraction of the Universe, whereby the frequency spectrum of light emitted by a cosmic body A differs from that detected by an observer placed in another cosmic body B. This phenomenon should not be confused with the **Doppler effect**, which also concerns the variation of the spectrum of some wave phenomenon (sound, light, etc.) emitted by a body A and received by another body B. In fact, the Doppler effect is due to the motion of a source with respect to an observer reference system.

6.1 Cosmic Redshift

The redshift phenomenon analyzed in this chapter considers two bodies A and B of the cosmic fluid, that is, two fixed points in the reference space, as shown in Fig. 6.1.

Let us consider two photons P and $\bar{P}$ emitted by A at two successive times $\bar{t}_{eA} > t_{eA}$ and intercepted by B at times $\bar{t}_{rB} > t_{rB}$. It is to be expected that the two emission-reception intervals have different amplitudes: $\bar{t}_{rB} - \bar{t}_{eA}$ and $t_{rB} - t_{eA}$.

Anyway, as shown in Fig. 6.2, by virtue of Remark 5.3 and Eq. (5.6) the two shaded areas above these two intervals remain unchanged because both are equal to $d_{AB}(t_\natural)$.

These shaded areas behave as if they were filled by a planar incompressible fluid channeled below the graph of $c/a(t, t_\natural)$ and above the emission-reception interval $[t_{eA}, t_{rB}]$.

Now suppose that the emission time $\bar{t}_{eA}$ of the second photon is very close to the emission time t_{eA} of the first photon, as shown in Fig. 6.3. The areas over the intervals $I_{er} = [t_{eA}, t_{rB}]$ and $\bar{I}_{er} = [\bar{t}_{eA}, \bar{t}_{rB}]$ are both equal to the distance $d_{AB}(t_\natural)$. Since the central white area is a common part of these two areas, the two shaded areas above the intervals $I_e = [t_{eA}, \bar{t}_{eA}]$ and $I_r = [t_{rB}, \bar{t}_{rB}]$ are equal.

If these two base-intervals are very small in comparison with the intervals I_{er} and $\bar{I}_{er}$ then the two shaded areas can be considered equal to the product *base* $\times$

S. Benenti, *Mathematical Foundations and Numerical Analysis of the Dynamics of an Isotropic Universe*, https://doi.org/10.1007/978-3-031-64033-9_6

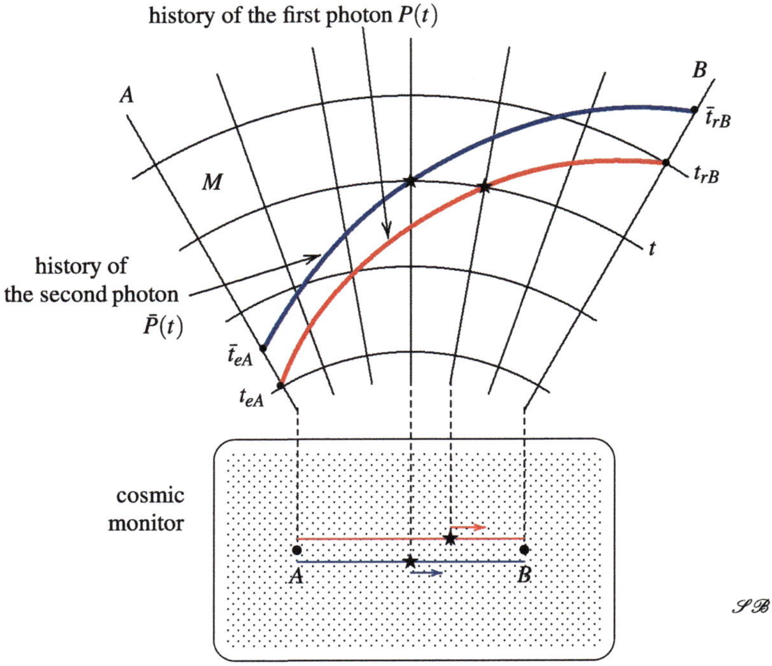

Fig. 6.1 Two photons P and $\bar{P}$ coming from A and received from B

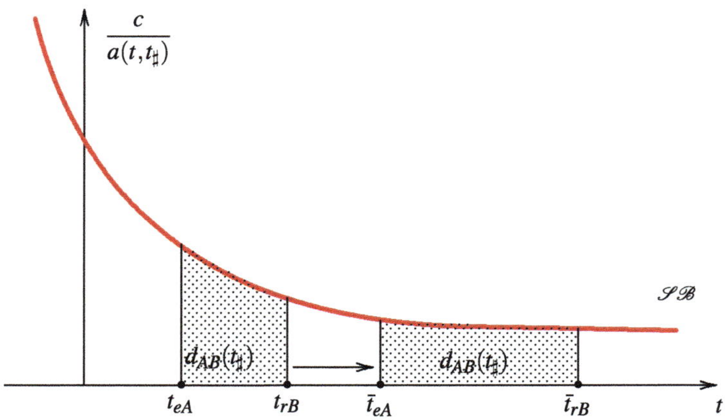

Fig. 6.2 Shifting of the emission-reception interval preserving the shaded area

height of the rectangles where they are contained. Thus, we can write with good approximation that[1]

---

[1] This argument is taken, with some modification, from [15], pp. 126–127.

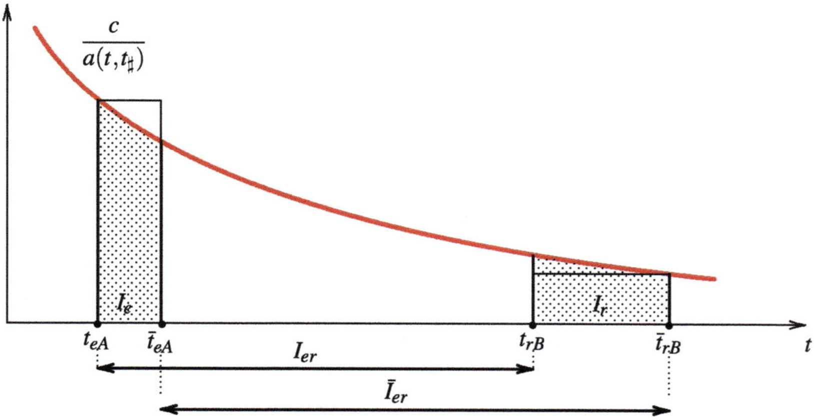

Fig. 6.3 The two shaded areas are equal

$$\frac{I_e}{a(t_{eA}, t_\sharp)} = \frac{I_r}{a(t_{rB}, t_\sharp)}$$

i.e.

$$\boxed{\frac{a(t_{rB}, t_\sharp)}{a(t_{eA}, t_\sharp)} = \frac{I_r}{I_e}} \tag{6.1}$$

This formula can be correctly applied when the emission events of the two photons correspond to two successive crests of a monochromatic wave of length

$$\lambda_{eA} = c\, I_e$$

emitted from A. Then the two receiving events from B correspond to two successive crests of the same wave but now of length

$$\lambda_{rB} = c\, I_r,$$

and (6.1) translates into equation

$$\boxed{\frac{a(t_{rB}, t_\sharp)}{a(t_{eA}, t_\sharp)} = \frac{\lambda_{rB}}{\lambda_{eA}}} \tag{6.2}$$

describing the phenomenon of **spectral shift** or **redshift**:

$$\begin{cases} a(t_{rB}, t_\sharp) > a(t_{eA}, t_\sharp) \iff \lambda_{rB} > \lambda_{eA} \iff \\ a(t_{rB}, t_\sharp) < a(t_{eA}, t_\sharp) \iff \lambda_{rB} < \lambda_{eA} \iff \end{cases}$$

$$\begin{cases} \iff \text{Shift of the original wavelength toward red.} \\ \iff \text{Shift of the original wavelength toward blue.} \end{cases}$$

Remark 6.1 If we write the (6.2) in the form

$$\lambda_{rB} = \frac{a(t_{rB}, t_\sharp)}{a(t_{eA}, t_\sharp)} \lambda_{eA}$$

we observe that the spectrum of a galaxy A observed from B is in fact *multiplied* by the ratio $a(t_{rB}, t_\sharp)/a(t_{eA}, t_\sharp)$ and not shifted, as the term 'shift' might suggest. •

By introducing the **redshift parameter**

$$\boxed{z_{AB} \overset{\text{def}}{=} \frac{\lambda_{rB} - \lambda_{eA}}{\lambda_{eA}} = \frac{\lambda_{rB}}{\lambda_{eA}} - 1} \tag{6.3}$$

Equation (6.2) takes the form

$$\boxed{\frac{a(t_{rB}, t_\sharp)}{a(t_{eA}, t_\sharp)} = 1 + z_{AB}} \tag{6.4}$$

We observe that this equation is $t_\sharp$-invariant, so that we can chose $t_\sharp = t_0$ and write

$$\boxed{\frac{a(t_{rB}, t_0)}{a(t_{eA}, t_0)} = 1 + z_{AB}} \tag{6.5}$$

The following theorem expresses the fundamental role played by redshift.

Theorem 6.1 *Equation*

$$\boxed{\frac{1}{a(t_{eA}, t_0)} = 1 + z_{AB}} \tag{6.6}$$

gives the relationship between the emission time t_{eA} of a photon emitted from A and the redshift z_{AB} measured today by an observer living in B.

Proof This equation is obtained from (6.5) by considering the reception time to be equal to the present time, $t_\sharp = t_0$, and then placing $a(t_0, t_0) = 1$. ■

Remark 6.2 For a generic time t Eq. (6.6) can be written in the simplified form

$$\boxed{\frac{1}{a(t, t_0)} = 1 + z} \tag{6.7}$$

Since $a(t, t_0)$ is an increasing function with no stationary points, thus invertible, *the redshift z can play the role of time parameter instead of t* as is customary in astronomy. Note that $z = 0$ corresponds to present time: $t = t_0$. •

Remark 6.3 Again for the same reason, Eq. (6.6) can be solved with respect to t_{eA}. This gives a function

$$z_{AB} \longmapsto t_{eA}$$

which, in accordance with Theorem 6.1, gives the emission time t_{eA} as a function of the redshift z_{AB} measured today by an observer in B. Actually, the emission time t_{eA} is equal to the age of the observed cosmic object A. In other words, *by measuring the redshift of a cosmic object we can calculate its age*, provided the analytical expression of $a(t, t_0)$ is known, as in the case of the MR-model. •

Theorem 6.2 *In the MR-model equation*

$$\boxed{t_{eA}(z_{AB}) = \frac{1}{\beta}\, arccosh\left(\frac{1}{\alpha^3\,(1+z_{AB}^3)}+1\right)} \tag{6.8}$$

gives the emission time of a signal emitted by A and picked up today by B with a redshift z_{AB}.

Proof Equation (6.8) is the inverse relation of (6.6), thus[2]:

$$(6.7)\ \ \frac{1}{a(t,t_0)} = 1+z \iff \frac{1}{a^3(t,t_0)} = (1+z)^3, \quad \text{profile (4.31)}$$

$$\Longrightarrow \frac{1}{\cosh(\beta\, t)-1} = \alpha^3\,(1+z)^3 \Longrightarrow \cosh(\beta\, t) - 1 = \frac{1}{\alpha\,(1+z)^3} \Longrightarrow (6.8).\ \blacksquare$$

Remark 6.4 Since $\operatorname{arccosh}(u) = \log(u+\sqrt{u^2-1})$, an alternative representation of $z_{AB} \mapsto t_{eA}$ is

$$\boxed{t_{eA} = \frac{1}{\beta}\,\log(u+\sqrt{u^2-1}), \quad u \overset{\text{def}}{=} \frac{1}{\alpha^3\,(1+z_{AB})^3}+1} \tag{6.9}$$

Here u plays the role of **intermediate parameter**. •

In Fig. 6.4, Sect. 5.2, the graph of the function $z_{AB} \mapsto t_{eA}$ is plotted. Some of its numerical values are given in Table 6.1 below (Sect. 5.3).

Remark 6.5 The reionization time t_{re} (beginning of light emission) corresponds to the maximum value of an observable redshift: $z_{\text{re}} = 8.8$, see Table 4.1. •

Remark 6.6 Calculation of the slope of the tangent at the initial point. The derivative of $y = \operatorname{arccosh}(x)$ is $y' = 1/\sqrt{x^2-1}$. Then from (6.8) we get

$$\beta\, t'(z) = \frac{1}{\sqrt{u^2-1}}\, u'(z).$$

---

[2] To simplify the calculation, let us put $z_{AB} = z$ e $t_{eA} = t$.

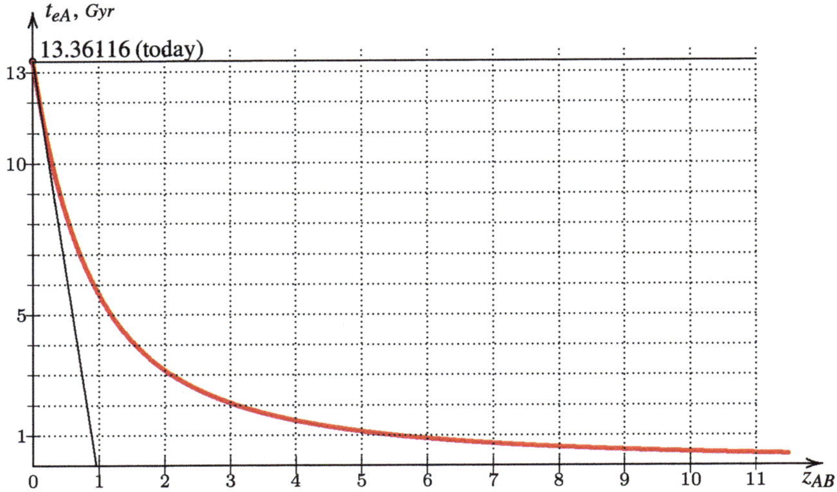

Fig. 6.4 $t_{eA}(z_{AB})$ in the MR-model

For $z = 0$ we have $u(0) = \dfrac{1+\alpha^3}{\alpha^3}, \quad \alpha^3\, u'(0) = -3$. It follows that

$$u^2(0) - 1 = \frac{(1+\alpha^3)^2}{\alpha^6} - 1 = \frac{(1+\alpha^3)^2 - \alpha^6}{\alpha^6} = \frac{1+2\alpha^3}{\alpha^6},$$

$$\beta\, t'(0) = \frac{\alpha^3}{\sqrt{1+2\alpha^3}}\, u'(0) = -\frac{3}{\sqrt{1+2\alpha^3}},$$

$$t'(0) = -\frac{3}{\beta\sqrt{1+2\alpha^3}}. \tag{6.10}$$

For $\alpha \simeq 0.607247$ and $\beta \simeq 0.178366\ Gyr^{-1}$ we get

$$t'(0) \simeq -13.978027. \tag{6.11}$$

This is the slope of the curve $t(z)$ at $z = 0$. The intersection of the tangent with the z-axis is located at $z \simeq 0.95586$. •

Table 6.1 $z_{AB} \mapsto t_{eA} \mapsto d_{AB}(t_0) \mapsto d_{AB}(t_{eA})$

z_{AB} today's redshift	t_{eA} emission time	$d_{AB}(t_0)$ today's distance	$d_{AB}(t_{eA})$ emission time distance
0.0	13.361160	0	0
0.1	12.059862	1.364801	1.240728
0.2	10.930753	2.661961	2.218301
0.3	9.947120	3.890388	2.992606
0.4	9.087011	5.050586	3.607562
0.5	8.332161	6.144310	4.096206
0.6	7.667264	7.174207	4.483879
0.7	7.079443	8.143518	4.790304
0.8	6.557845	9.055803	5.031001
0.9	6.093308	9.914755	5.218292
1.0	5.678087	10.724054	5.362027
1.2	4.970357	12.207806	5.549003
1.4	4.393193	13.533387	5.638911
1.6	3.916490	14.723675	5.662952
1.8	3.518136	15.798076	5.642170
2.0	3.181690	16.772852	5.590950
2.2	2.894767	17.661574	5.519241
2.4	2.647919	18.475571	5.433991
2.6	2.433848	19.224324	5.340090
2.8	2.246851	19.915803	5.241000
3.0	2.082421	20.556736	5.139184
3.2	1.936959	21.152840	5.036390
3.4	1.807564	21.708992	4.933861
3.6	1.691877	22.229372	4.832472
3.8	1.587960	22.717600	4.732833
4.0	1.494213	23.176803	4.635360
4.2	1.409303	23.609706	4.540328
4.4	1.332111	24.018703	4.447908
4.6	1.261694	24.405891	4.358194
4.8	1.197251	24.773123	4.271228
5.0	1.138098	25.122042	4.187007
5.2	1.083649	25.454107	4.105501
5.4	1.033398	25.770622	4.026659
5.6	0.986906	26.072761	3.950418
5.8	0.943791	26.361578	3.876702
6.0	0.903722	26.638006	3.805429
6.2	0.866406	26.902906	3.736514
6.4	0.831585	27.157060	3.669873

(continued)

Table 6.1 (continued)

z_{AB} today's redshift	t_{eA} emission time	$d_{AB}(t_0)$ today's distance	$d_{AB}(t_{eA})$ emission time distance
6.6	0.799032	27.401171	3.605417
6.8	0.768547	27.635873	3.543060
7.0	0.739949	27.861767	3.482720
7.2	0.713081	28.079370	3.424313
7.4	0.687798	28.289194	3.367761
7.6	0.663973	28.491683	3.312986
7.8	0.641492	28.687246	3.259914
8.0	0.620250	28.876280	3.208475
8.2	0.600154	29.059135	3.158601
8.4	0.581120	29.236134	3.110227
8.6	0.563070	29.407594	3.063291
8.8	0.545936	29.573779	3.017732

6.2 From Redshift to the Current Distance

Theorem 6.3 *If at the present time t_0 a cosmic body A is observed from B with a redshift $z = z_{AB}$ then the current distance $d_{AB}(t_0)$, measured in the spatial section S_{t_0}, is given by the integral*

$$\boxed{d_{AB}(t_0) = c \int_{t(z)}^{t_0} \frac{dt}{a(t, t_0)}} \tag{6.12}$$

where the lower bound of integration $t(z) = t_{eA}(z_{AB})$ is given by the (6.8).

Proof We put $t_{eA} = t(z)$, $t_\sharp = t_0$ and $t_{rB} = t_0$ in the emission-reception relationship (5.2). ∎

The graph of the distance (6.12) is plotted in Fig. 6.5, where it is compared with the reception distance $d_{AB}(t_{eA})$ given by Eq. (6.13) in the next section. See also Table 6.1 on Sect. 5.3 for some numerical values (Fig. 6.6).

6.3 From Redshift to the Distance at the Emission Time

Theorem 6.4 *If at the present time t_0 a cosmic object A is observed from B with a redshift $z = z_{AB}$ then the distance $d_{AB}(t(z))$ measured in the spatial section corresponding to the emission time is given by*

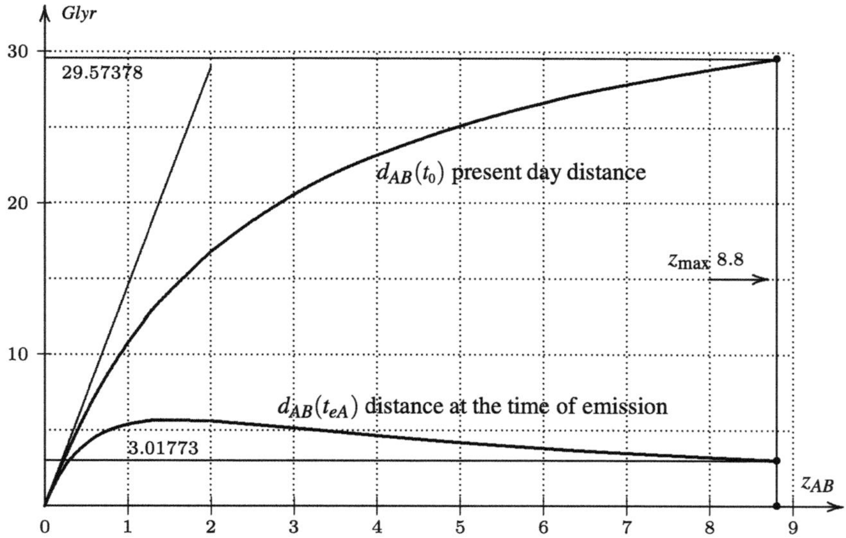

Fig. 6.5 Redshift z_{AB} versus present day distances and emission time

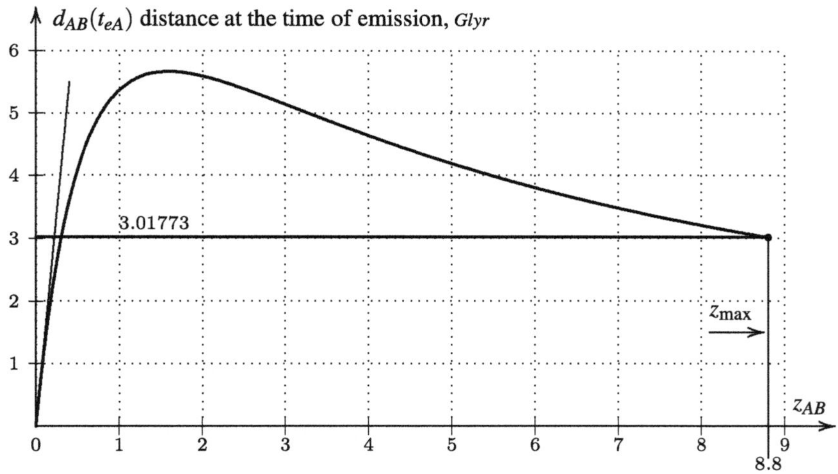

Fig. 6.6 Redshift z_{AB} versus distance $d_{AB}(t_{eA})$ at the time of emission

$$\boxed{d_{AB}(t_{eA}) = \frac{c}{1+z_{AB}} \int_{t(z)}^{t_0} \frac{dt}{a(t, t_0)} = \frac{1}{1+z_{AB}}\, d_{AB}(t_0)} \tag{6.13}$$

Proof From formula (1.30) written with $t_\sharp = t_0$, $d_{AB}(t) = a(t, t_0)\, d_{AB}(t_0)$, as well as from (6.12), it follows

$$d_{AB}(t) = c\, a(t, t_0) \int_{t(z)}^{t_0} \frac{dt}{a(t, t_0)}.$$

Posing $t = t(z)$ results in the following:

$$d_{AB}(t(z)) = c\ a(t(z), t_0) \int_{t(z)}^{t_0} \frac{dt}{a(t, t_0)}.$$

Finally, we apply (6.7) $\dfrac{1}{a(t, t_0)} = 1 + z$. ■

On each row of Table 6.1 we find the redshift z_{AB} of a cosmic body A measured today by B, the emission time t_{eA}, the present day distance $d_{AB}(t_0)$ and the distance at the emission time $d_{AB}(t_{eA})$.

- The mapping $z_{AB} \mapsto t_{eA}$ is defined by Eq. (6.9)

$$\boxed{t_{eA} = \frac{1}{\beta} \log(u + \sqrt{u^2 - 1}), \quad u \stackrel{\text{def}}{=} \frac{1}{\alpha^3\,(1 + z_{AB})^3} + 1}$$

hence by

$$\boxed{t_{eA}(z_{AB}) = \frac{1}{\beta} \operatorname{arccosh}\left(\frac{1}{\alpha^3\,(1 + z_{AB}^3)} + 1\right)} \tag{6.14}$$

- The mapping $t_{eA} \mapsto d_{AB}(t_0)$ is defined by Eq. (6.12)

$$d_{AB}(t_0) = c \int_{t_{eA}}^{t_0} \frac{dt}{a(t, t_0)}.$$

- The mapping $d_{AB}(t_0) \mapsto d_{AB}(t_{eA})$ is defined by Eq. (6.13)

$$d_{AB}(t_{eA}) = \frac{1}{1 + z_{AB}}\, d_{AB}(t_0).$$

Remark 6.7 The distance of 29 billion light-years that we see in the last rows (middle column) of the table may seem incredibly high in a Universe that is 'only' 13.36 billion years old, where a light-year is the distance light travels in one year and where nothing can travel faster than light. This is a paradox that has generated many misunderstandings about the size of the visible Universe and the measurement of distances in an expanding space. •

6.4 Slipher Law and Constant Expansion Models

According to [11], p. 274: "*Howard Robertson in 1928 showed that Slipher's redshifts and Hubble's previously published instances supported an approximate redshift–distance relation*

$$z\,c = H L \quad [14.5]$$

where L is the distance of the galaxy, and the constant H is called Hubble term".

In our context, equation [14.5] translates into **Slipher's law**

$$\boxed{d_{AB}(t_0, z) = r_H\, z} \tag{6.15}$$

where

$$r_H \overset{\text{def}}{=} \frac{c}{H_0}$$

is the Hubble radius (4.52). This law gives the current distance of a galaxy A in terms of the redshift z measured from B.

Today we know that the Slipher law (6.15) is acceptable only for small values of z. However, it is worthwhile to analyze the mathematical aspects underlying this law. We will see that it is the result of two successive linearization processes.

A **constant-rate expansion model** (CREM) is based on the assumption that the Universe has evolved with a constant growth rate throughout its history. Such a model is then characterized by a scale factor of the type

$$a_{\text{cre}}(t, t_\sharp) = \mu\,(t - t_\sharp) + \nu, \quad \mu,\ \nu = \text{constant}, \ \ \mu > 0.$$

However, the normalization condition $a_{\text{cre}}(t_\sharp, t_\sharp) = 1$ implies $\nu = 1$ so the previous formula reduces to

$$a_{\text{cre}}(t, t_\sharp) = \mu\,(t - t_\sharp) + 1.$$

The corresponding Hubble factor is

$$H(t) = \frac{\dot{a}_{\text{cre}}}{a_{\text{cre}}} = \frac{\mu}{\mu\,(t - t_\sharp) + 1}.$$

By setting $t = t_\sharp$ we find $H(t_\sharp) = \mu > 0$, so ***the profile of a constant expansion model turns out to be***

$$\boxed{a_{\text{cre}}(t, t_\sharp) = H(t_\sharp)\,(t - t_\sharp) + 1} \tag{6.16}$$

Hence we observe that ***a constant expansion model is uniquely determined by the $H(t_\sharp)$ value of the Hubble factor at a given reference time.***

In the plane (t, a) the profile (6.16) is a line through the point $(t_\sharp, 1)$ with slope $H(t_\sharp)$, Fig. 6.7. This line intersects the t-axis at a point t_α marking the start date of the Universe. By setting $t = t_\alpha$ and $a_{\text{cre}}(t_\alpha, t_\sharp) = 0$ in Eq. (6.16) we find

$$\boxed{t_\alpha = t_\sharp - \frac{1}{H(t_\sharp)}} \tag{6.17}$$

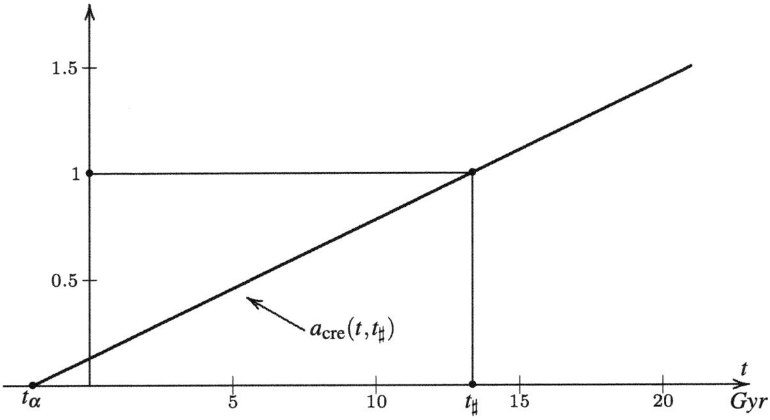

Fig. 6.7 Constant-expansion model profile

As Fig. 6.7 shows, this date can have a negative value. This means that the CRE-model predicts a birth date of the Universe earlier than the date of the big-bang (see below).

Theorem 6.5 *At the point* $(t_\sharp, 1)$ *the line* $a_{\text{cre}}(t, t_\sharp)$ *is tangent to each profile* $a(t, t_\sharp)$ *satisfying equation*

$$\dot{a}(t_\sharp, t_\sharp) = H(t_\sharp). \tag{6.18}$$

Proof From Eq. (6.16) it follows $\dot{a}_{\text{cre}}(t, t_\sharp) = H(t_\sharp)$ for each t and thus $\dot{a}_{\text{cre}}(t_\sharp, t_\sharp) = H(t_\sharp)$. For every other profile $a(t, t_\sharp)$ with reference time $t_\sharp$ the same equation $\dot{a}(t_\sharp, t_\sharp) = H(t_\sharp)$ holds. ■

In Fig. 6.8 the profile $a(t, t_0)$ (4.38) of the MR-model is compared with the profile of the CRE-model with $t_\sharp = t_0$.

Since $H(t_\sharp) = H_0 \simeq 0.07154\ Gyr^{-1}$ e $t_0 \simeq= 13.36116\ Gyr$, the numerical expression of the profile (6.16) is

$$\boxed{a_{\text{cre}}(t, t_0) = H_0\,(t - t_0) + 1 \simeq 0.07154 * (t - 13.36116) + 1} \tag{6.19}$$

By virtue of the Theorem 6.5 the two profiles are tangent at the point $(t_0, 1)$. Applying (6.17) we find

$$t_\alpha = t_0 - \frac{1}{H_0} = t_0 - t_H$$

where

$$\boxed{t_H = H_0^{-1} \simeq 13.97802\,Gyr} \tag{6.20}$$

is the so-called **Hubble time**. It follows that the beginning of the Universe is dated.

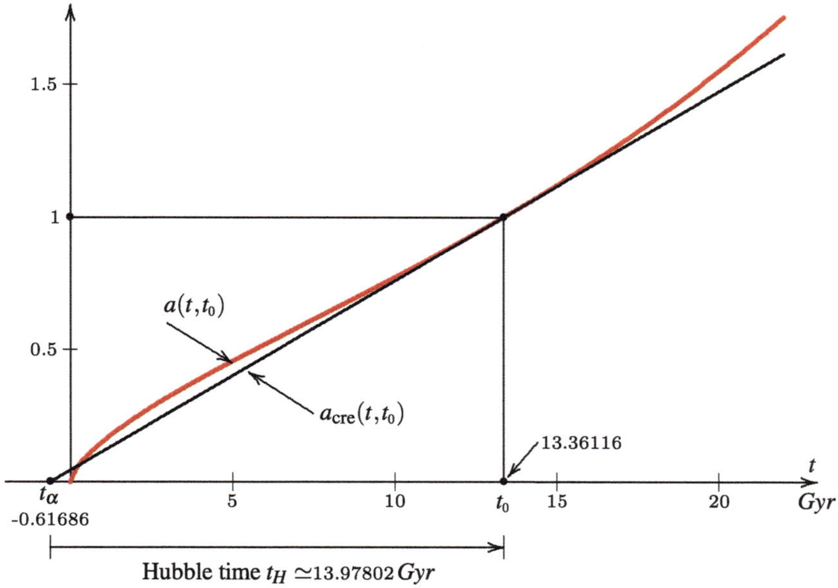

Fig. 6.8 Profile of the constant expansion model compared with the MR-model

$$\boxed{t_\alpha \simeq -0.61686\ Gyr} \tag{6.21}$$

Let us now examine the transmission of photons in the CRE model

Theorem 6.6 *In a constant expansion model, a photon emitted from A at time t_{eA} reaches B at time t_{rB} if and only if*

$$\boxed{d_{AB}(t_\sharp) = \frac{c}{H(t_\sharp)} \log \frac{t_{rB} - t_\alpha}{t_{eA} - t_\alpha}} \tag{6.22}$$

Proof The emission-reception relationship (5.2) is valid for every profile and is independent of the choice of reference time (Theorem 5.4). Applying it to the profile $a_{\text{cre}}(x, t_\sharp)$ we get

$$c \int_{t_e}^{t_r} \frac{dx}{a_{\text{cre}}(x, t_\sharp)} = d_{AB}(t_\sharp). \tag{6.23}$$

It follows that

$$\int_{t_e}^{t_r} \frac{dx}{a_{\text{cre}}(x, t_\sharp)} = \int_{t_e}^{t_r} \frac{dx}{H(t_\sharp)(x - t_\sharp) + 1} = \frac{1}{H(t_\sharp)} \Big[\log \big[H(t_\sharp)(x - t_\sharp) + 1 \big] \Big]_{t_e}^{t_r}$$

$$= \frac{1}{H(t_\sharp)} \Big[\log \big[H(t_\sharp)(t_r - t_\sharp) + 1 \big] - \log \big[H(t_\sharp)(t_e - t_\sharp) + 1 \big] \Big]$$

$$= \frac{1}{H(t_\sharp)} \log \frac{H(t_\sharp)(t_r - t_\sharp) + 1}{H(t_\sharp)(t_e - t_\sharp) + 1} = \dots$$

Let us apply (6.17) $\dfrac{1}{H(t_\sharp)} = t_\sharp - t_\alpha$,

$$\dots = \frac{1}{H(t_\sharp)} \log \frac{(t_r - t_\sharp) + (t_\sharp - t_\alpha)}{(t_e - t_\sharp) + (t_\sharp - t_\alpha)} = \frac{1}{H(t_\sharp)} \log \frac{t_r - t_\alpha}{t_e - t_\alpha}. \qquad \blacksquare$$

Theorem 6.7 *In a constant expansion model, if B observes A with a redshift z_{AB} then the isochronous distance between A and B at time $t_\sharp$ is given by*

$$\boxed{d_{AB}(t_\sharp, z_{AB}) = \frac{c}{H(t_\sharp)} \log(1 + z_{AB})} \tag{6.24}$$

Proof The general Eq. (6.4) applied to a_{cre}; translates to

$$\frac{H(t_\sharp)(t_r - t_\sharp) + 1}{H(t_\sharp)(t_e - t_\sharp) + 1} = 1 + z.$$

Due to (6.17), $H^{-1}(t_\sharp) = t_\sharp - t_\alpha$, we find $1 + z = \dfrac{t_r - t_\sharp + H^{-1}(t_\sharp)}{t_e - t_\sharp + H^{-1}(t_\sharp)} = \dfrac{t_r - t_\alpha}{t_e - t_\alpha}$.

Hence (6.22) $\Longrightarrow$ (6.24). $\blacksquare$

In particular, for $t_\sharp = t_0$ from (6.24) it follows

$$\boxed{d_{AB}(t_0, z) = r_H \log(1 + z_{AB})} \tag{6.25}$$

The graphs of (6.25) and (6.12) are compared in Fig. 6.9. They have the same tangent line at the origin with slope $\simeq 13.97802$.

About formula (6.25) it should be noted that the logarithmic series

$$\log(1 + z) = z - \frac{z^2}{2} + \frac{z^3}{3} - \frac{z^4}{4} + \frac{z^5}{5} + \cdots$$

converges very slowly since the graphs of its reduced sums

$$z, \quad z - \frac{z^2}{2}, \quad z - \frac{z^2}{2} + \frac{z^3}{3}, \quad z - \frac{z^2}{2} + \frac{z^3}{3} - \frac{z^4}{4}, \dots$$

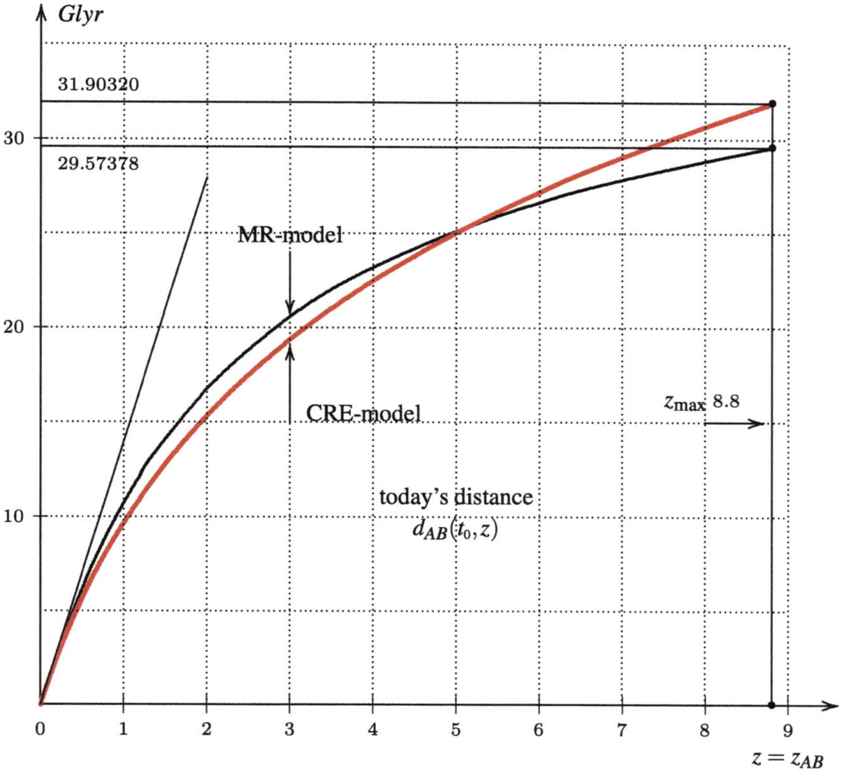

Fig. 6.9 $z_{AB} \mapsto d_{AB}(t_0, z)$, comparison between the MR-model and the CRE-model

differ significantly. Consequently, the only approximation that makes sense is

$$\log(1 + z) \simeq z$$

and this is acceptable only for small values of z, in full agreement with the Slipher law (6.15).

Chapter 7
Appendices

7.1 Stereographic Projection of Hyper-spheres

Equation

$$X_1^2 + \cdots + X_{n+1}^2 = r^2. \tag{7.1}$$

defines a hyper-sphere $\mathbb{S}_n \subset \mathbb{R}^{n+1} = (X_1, X_2, \ldots, X_{n+1})$ with radius r and centered at the origin $O = (0, \ldots, 0)$. Figure 7.1 represents the stereographic projection from the North pole $N = (0, 0, \ldots, r)$ onto the Cartesian plane $\mathbb{R}^n = (x_i) = (x_1, ..., x_n)$ tangent to the South pole $S = (0, \ldots, -r)$. A generic point $B = [X_i(B), X_{n+1}(B)]$ di $\mathbb{S}_n$ is projected to the point $C = [x_i(C)]$ of the plane.

Theorem 7.1 *The stereographic projection $\mathbb{S}_n \mapsto \mathbb{R}^n$ provides the following parametric representation of $\mathbb{S}_n$*

$$\boxed{\begin{aligned} X_i &= \frac{d^2}{\rho^2 + d^2}\, x_i, \\ X_{n+1} &= \frac{\rho^2 - d^2}{\rho^2 + d^2}\, r, \end{aligned} \qquad d \stackrel{\text{def}}{=} 2r, \quad \rho^2 \stackrel{\text{def}}{=} \sum_{i=1}^{n} x_i^2} \tag{7.2}$$

with parameters $x_i \stackrel{\text{def}}{=} x_i(C)$.

Note that the North pole N is excluded from this representation.
Proof Because the similarity of the triangles NAB and NSC we have

$$\frac{BA}{CS} = \frac{NA}{NS}. \tag{7.3}$$

S. Benenti, *Mathematical Foundations and Numerical Analysis of the Dynamics of an Isotropic Universe*, https://doi.org/10.1007/978-3-031-64033-9_7

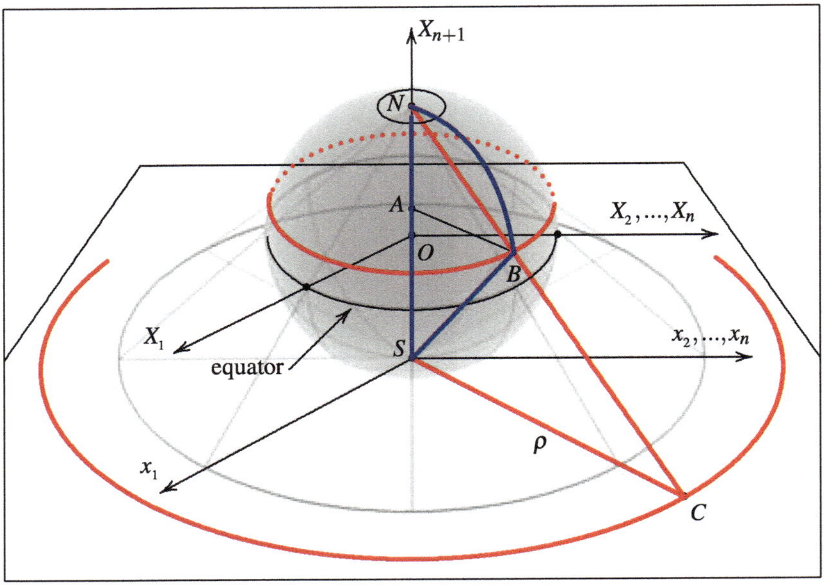

Fig. 7.1 Stereographic projection of $\mathbb{S}_n$: $N \to B \to C$

Since

$$\begin{cases} NA = r - X_{n+1}(A) = r - X_{n+1}(B), \\ NS = 2r = d \quad \text{(diameter of the hyper-sphere)}, \end{cases}$$

and

$$\frac{BA}{CS} = \frac{X_i(B)}{x_i(C)}, \quad i = 1, \ldots, n,$$

equation (7.3) is equivalent to

$$\alpha \overset{\text{def}}{=} \frac{X_i(B)}{x_i(C)} = \frac{r - X_{n+1}(B)}{d} \tag{7.4}$$

whatever index i. It follows that

$$\begin{cases} X_i(B) = \alpha\, x_i(C). \\ r - X_{n+1}(B) = \alpha\, d \implies X_{n+1}(B) = r - \alpha\, d. \end{cases} \tag{7.5}$$

Since point B lies on the sphere, its components satisfy equation

$$\sum_{i=1}^{n} X_i^2(B) + X_{n+1}^2(B) = r^2.$$

Due to (7.5) this equation becomes

$$\alpha^2 \sum_{i=1}^{n} X_i^2(C) + (r - \alpha\, d)^2 = r^2. \tag{7.6}$$

Since $X_i(C) = x_i(C)$, by posing

$$x_i \stackrel{\text{def}}{=} x_i(C), \quad \rho^2 \stackrel{\text{def}}{=} \sum_{i=1}^{n} x_i^2,$$

we have the sequence

$$(7.6) \iff \alpha^2\, \rho^2 + (r - \alpha\, d)^2 = R^2 \iff \alpha^2\, (\rho^2 + d^2) - 2\,\alpha\, r\, d = 0$$

$$\iff \alpha\, (\rho^2 + d^2) = d^2 \iff \alpha = \frac{d^2}{\rho^2 + d^2}.$$

Then from (7.5) we get

$$\begin{cases} X_i(B) = \alpha\, x_i = \dfrac{d^2}{\rho^2 + d^2}\, x_i. \\ X_{n+1}(B) = r - \alpha\, d = r - \dfrac{d^2}{\rho^2 + d^2}\, d = r \left(1 - 2\, \dfrac{d^2}{\rho^2 + d^2}\right) = \dfrac{\rho^2 - d^2}{\rho^2 + d^2}\, r. \end{cases} \quad \blacksquare$$

Theorem 7.2 *The holonomic reference frame associated with the coordinates* (x_i) *consists of the n independent vectors tangent to* $\mathbb{S}_n$ *defined as follows:*

$$\boldsymbol{E}_i \stackrel{\text{def}}{=} \begin{bmatrix} \dfrac{\partial X_j}{\partial x_i} \\ \dfrac{\partial X_{n+1}}{\partial x_i} \end{bmatrix} = \frac{d^2}{(\rho^2 + d^2)^2} \begin{bmatrix} (\rho^2 + d^2)\, \delta_{ij} - 2\, x_i\, x_j \\ 4\, r\, x_i \end{bmatrix}. \tag{7.7}$$

Proof[A] : $\dfrac{\partial}{\partial x_i} \dfrac{1}{\rho^2 + d^2} = -2\, (\rho^2 + d^2)^{-2}\, x_i.$

[B] : $\left[\dfrac{\partial}{\partial x_i} \dfrac{\rho^2 - d^2}{\rho^2 + d^2} = \dfrac{2\, x_i\, (\rho^2 + d^2) - (\rho^2 - d^2)\, 2\, x_i}{(\rho^2 + d^2)^2} = \dfrac{4\, d^2}{(\rho^2 + d^2)^2}\, x_i.\right.$

$$[\text{A}] \Longrightarrow \left[\begin{aligned} &\frac{\partial X_j}{\partial x_i} = x_j\, \frac{\partial}{\partial x_i} \frac{d^2}{\rho^2 + d^2} + \frac{d^2}{\rho^2 + d^2}\, \delta_{ij} \\ &= d^2 \left[-2\, (\rho^2 + d^2)^{-2}\, x_i\, x_j + (\rho^2 + d^2)^{-1}\, \delta_{ij}\right] \\ &= d^2\, (\rho^2 + d^2)^{-2} \left[(\rho^2 + d^2)\, \delta_{ij} - 2\, x_i\, x_j\right] \quad [\text{C}]. \end{aligned}\right.$$

$$[\mathrm{B}] \Longrightarrow \left[\frac{\partial X_{n+1}}{\partial x_i} = r\,\frac{\partial}{\partial x_i}\frac{\rho^2 - d^2}{\rho^2 + d^2} = 4\,r\,d^2\,(\rho^2 + d^2)^{-2}\,x_i \quad [\mathrm{D}]. \right.$$

[C]+[D] $\Longrightarrow$ (7.7). ■

Theorem 7.3 *The canonical Euclidean metric of* $\mathbb{R}^{n+1}$ *induces a Riemannian metric on* $\mathbb{S}_n$ *whose* g_{ij} *components in the* (x_i) *coordinates are*

$$\boxed{g_{ij} \stackrel{\text{def}}{=} \boldsymbol{E}_i \cdot \boldsymbol{E}_j = \left(1 + \frac{1}{4}\,\frac{\sum_i x_i^2}{r^2}\right)^{-2} \delta_{ij}} \tag{7.8}$$

Proof

$$\left[\begin{aligned} & g_{ij} \stackrel{\text{def}}{=} \boldsymbol{E}_i \cdot \boldsymbol{E}_j = \\ & \frac{d^4}{(\rho^2 + d^2)^4} \sum_k \begin{bmatrix} (\rho^2 + d^2)\,\delta_{ik} - 2\,x_i\,x_k \\ 4\,r\,x_i \end{bmatrix} \cdot \begin{bmatrix} (\rho^2 + d^2)\,\delta_{kj} - 2\,x_k\,x_j \\ 4\,r\,x_j \end{bmatrix} \\ & = \frac{d^4}{(\rho^2 + d^2)^4} \Big((\rho^2 + d^2)^2\,\delta_{ij} - 4\,(\rho^2 + d^2)\,x_i\,x_j + 4\,\rho^2\,x_i\,x_j + 16\,r^2\,x_i\,x_j \Big) \\ & = \frac{d^4}{(\rho^2 + d^2)^4} \Big((\rho^2 + d^2)^2\,\delta_{ij} - 4\,d^2\,x_i\,x_j + 16\,r^2\,x_i\,x_j \Big) = \frac{d^4}{(\rho^2 + d^2)^2}\,\delta_{ij}. \end{aligned} \right.$$

Since

$$\frac{d^4}{(\rho^2 + d^2)^2} = \left(1 + \frac{\rho^2}{d^2}\right)^{-2}$$

we find (7.8). ■

Remark 7.1 Comparison of (7.8) with (1.19) shows that $\mathbb{S}_n$ is a manifold with constant curvature $K = 1/r^2$ and that the parametric coordinates (x_i) are curvature coordinates. •

Remark 7.2 A geodesic of $\mathbb{S}_3$ is a maximal circle, that is, the intersection of the sphere with a 3-plane passing through the origin. We say that a geodesic arc of amplitude ψ (angle to the center of the sphere) has length equal to $\psi\, R$. Thus the length of a maximal circle is $2\,\psi\, R$ (Fig. 7.2). •

7.2 Volume of a Hyper-spherical Cap

The intersection of the hyper-sphere $\mathbb{S}_3$ of radius r

$$\textstyle\sum_a x_a^2 + x_4^2 = r^2, \quad a = 1, 2, 3,$$

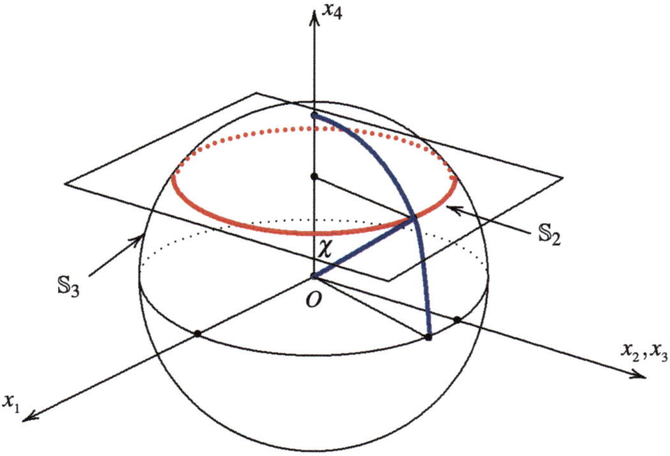

Fig. 7.2 Volume of a hyper-spherical cap: $\mathbb{S}_2 \subset \mathbb{S}_3 \subset \mathbb{R}^4$

with the hyper-plane $x_4 = r\,\cos\chi,\ \chi \in [0, \pi]$, is the sphere $\mathbb{S}_2 \subset \mathbb{R}^3 = (x_a)$

$$\textstyle\sum_a x_a^2 = r^2\,\sin^2\chi$$

of radius $r\,\sin\chi$. In turns, the hyper-sphere $\mathbb{S}_3$ admits a parametric representation of parameters (χ, u, v)

$$\begin{cases} x_a = r\,\sin\chi\,\xi_a(u, v), \\ x_4 = r\,\cos\chi, \end{cases}$$

where $x_a = \xi_a(u, v)$ is any parametric representation of the sphere of unitary radius $\sum_a \xi_a^2 = 1$. Since

$$\begin{cases} dx_a = r\left(\cos\chi\,\xi_a\,d\chi + \sin\chi\,d\xi_a\right), \\ dx_4 = -r\,\sin\chi\,d\chi, \end{cases}$$

the square of the arc-element on $\mathbb{S}_3$ is

$$\begin{cases} dS^2 = \sum_a dx_a^2 + dx_4^2 = r^2\left(\cos^2\chi\,d\chi^2\,\sum_a \xi_a^2 + \sin^2\chi\,\sum_a d\xi_a^2\right) + r^2\,\sin^2\chi\,d\chi^2 \\ = r^2\left(\cos^2\chi\,d\chi^2 + \sin^2\chi\,\sum_a d\xi_a^2\right) + r^2\,\sin^2\chi\,d\chi^2, \end{cases}$$

$$\boxed{dS^2 = r^2\,(d\chi^2 + \sin^2\chi\,d\sigma^2)}$$

where

$$d\sigma^2 = \sum_a d\xi_a^2$$

is the square of the arc-element on the unitary sphere. We can use spherical coordinates $(u, v) = (\theta, \phi)$ with $\theta \in [0, \pi]$ and $\phi \in [0, 2\pi]$, so

$$d\sigma^2 = d\theta^2 + \sin^2\theta \, d\phi^2$$

and the metric tensor on $\mathbb{S}_3$ turns out to be

$$[g_{ab}] = r^2 \begin{bmatrix} 1 & 0 & 0 \\ 0 & \sin^2\chi & 0 \\ 0 & 0 & \sin^2\chi \, \sin^2\theta \end{bmatrix}.$$

Hence $g \stackrel{\text{def}}{=} \det[g_{ij}] = r^6 \sin^4\chi \, \sin^2\theta$ end $\sqrt{g} = r^3 \sin^2\chi \, \sin\theta$, so that the volume-element is

$$dV = \sqrt{g} \, d\chi \, d\theta \, d\phi = r^3 \sin^2\chi \, \sin\theta \, d\chi \, d\theta \, d\phi.$$

Since

$$\int \sin^2 x \, dx = \tfrac{1}{2}\,(x - \sin x \, \cos x) + \text{constant},$$

the volume of the spherical cap of radius r and semi-amplitude ψ is given by the integral

$$V(r, \psi) = r^3 \int_0^{\psi} \sin^2\chi \, d\chi \cdot \int_0^{\pi} \sin\theta \, d\theta \cdot \int_0^{2\pi} d\phi = r^3 \, \tfrac{1}{2}\,\pi\,(\psi - \sin\psi \, \cos\psi) \cdot 2 \cdot 2\pi.$$

As a consequence (Fig. 7.3)

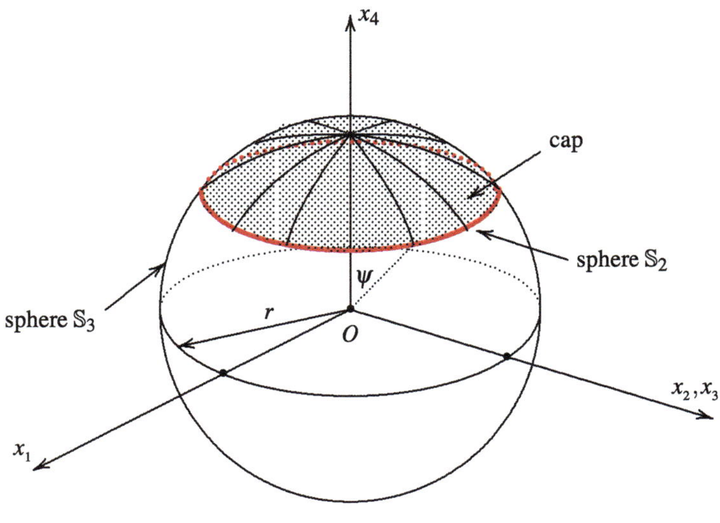

Fig. 7.3 Spherical cap of $\mathbb{S}_3$

$$\boxed{V(r, \psi) = 2\pi\, r^3\, (\psi - \sin\psi\ \cos\psi)} \tag{7.9}$$

By setting $\psi = \pi$ we obtain the volume of the whole sphere $\mathbb{S}_3$:

$$\boxed{V = 2\pi^2\, r^3} \tag{7.10}$$

Correction to: Numerical Cosmology

Correction to:
Chapter 4 in: S. Benenti, *Mathematical Foundations and Numerical Analysis of the Dynamics of an Isotropic Universe*, https://doi.org/10.1007/978-3-031-64033-9_4

The original version of the book was inadvertently published with the incorrect equations for 4.8 and 4.9, which have now been updated. the book has been updated with the changes.

The updated version of this chapter can be found at
https://doi.org/10.1007/978-3-031-64033-9_4

S. Benenti, *Mathematical Foundations and Numerical Analysis of the Dynamics of an Isotropic Universe*, https://doi.org/10.1007/978-3-031-64033-9_8

References

1. Astronomy & Astrophysics **594**, A11 (2016), *Planck 2015 results XI. CMB power spectra, likelihoods, and robustness of parameters*, Table 21 (six-parameter ΛCDM cosmology), two columns.
2. A&A **594**, A13 (2016), *Planck 2015 results XIII. Cosmological parameters*, Table 8 (ΛCDM cosmology), last column.
3. A&A **596**, A107 (2016), *Planck intermediate results XLVI. Reduction of large-scale etc.*, Table 8 (ΛCDM cosmology), last column.
4. Ade P.A.R. et al., *Planck 2015 results XIII. Cosmological parameters. A&A* **594** (A13), 1–63 (2016).
5. Basri, S.A. 1966. *A Deductive Theory of Space and Time, Studies in Logic and the Foundations of Mathematics*. Amsterdam: North-Holland Publishing Co.
6. Dodd R., *Using SI units in astronomy*, Cambridge University Press (2012).
7. L.P. Eisenhart, *Riemannian Geometry*, Princeton University Press (1925).
8. L.P. Eisenhart, *Non-Riemannian Geometry*, Am. Math. Society (1949).
9. Friedmann, A. 1999. Über die Krümmung des Raumes, Zeitschrift für Physik, 10 (1), 377–386 (1922). *English translation: On the Curvature of Space, General Relativity and Gravitation* 31 (12): 1991–2000.
10. Friedmann A., *Über die Möglichkeit einer Welt mit kostanter negativer Krümmung des Raumes*, Zeitschrift für Physik, **21**, 326–332 (1924). English translation: *On the Possibility of a World with Constant Negative Curvature of Space*, General Relativity and Gravitation, **31** (12), 2001–2008 (1999).
11. Harrison, E. 2012. *Cosmology, the Science of the Universe*. Press: Cambridge Univ.
12. Lemaître, G. 1927. Un Univers homogène de masse constante et de rayon croissant rendant compte de la vitesse radiale des nébuleuses extra-galactiques. *Annales de la Société Scientifique de Bruxelles* 47: 49–59.
13. Levi-Civita T., Amaldi U., *Lezioni di Meccanica Razionale* (1926, reprint 1989), volume II, part I, §6.
14. LIGO and Virgo Collaboration, *A gravitational-wave standard siren measurement of the Hubble constant*, Nature **551** (November 2017).
15. Liddle A., *An Introduction to Modern Cosmology*, Wiley 1998.
16. Lee, J.M. 1997. *Riemannian Manifolds*. An Introduction to Curvature: Springer.
17. Narlikar, J.L. 2002. *An Introduction to Cosmology*. Press: Cambridge Univ.

S. Benenti, *Mathematical Foundations and Numerical Analysis of the Dynamics of an Isotropic Universe*, https://doi.org/10.1007/978-3-031-64033-9

18. Oesch P.A. et al., *A remarkably luminous galaxy at* $z = 11.1$ *measured with Hubble space telescope grism spectroscopy*, The Astrophysical Journal, 819:129, 2016 March 10.
19. Olive, K.A., et al. 2014. Particle Data Group. *Chinese Physics C* 38: 090001.
20. Riess A.G. et al., *A 2.4% Determination of the Local Value of the Hubble Constant*, The Astrophysical Journal, Volume 826, Number 1, 2016.
21. *Wilkinson Microwave Anisotropy Probe (WMAP) project*, 7th year, 2010.
22. Wolf J.A., *Spaces of constant curvature*, Publish or Perish, Inc., 1984.

Index

A
Absolute
  invisibility radius, 109
  time, 34
  velocity, 45, 49
Absolute invisibility radius, xiv
Acceleration, vii
Acceleration vector, 7
Affine parameter, 7
Age of the Universe, xi, 64
Asymptotic objective, 63, 70
Auxiliary
  constant, 45
  velocity, 18

B
Balancing time, ix
Baryonic energy density, 55
Big-bang, xi, 66
Big-crunch, 66

C
Characteristic functions, 19
Christoffel
  symbols, 7
  symbols of first kind, 7
  symbols of second kind, 7
Chronology of events, 3
Closed Universe model, 80
Cold dark matter energy density, 55
Co-moving
  coordinates, 17
  distance, 14
Compatibility requirements, 22
Conformal transformation, 8
Congruence of curves, 1
Connection, vii, 5
Constant curvature, 8
Constant-rate expansion model (CREM), 125
Copernican principle, 3
Cosmic
  background radiation, 101
  body, 1
  connection, 22
  fluid, 1
  monitor, 20
  space-time, 1
  time, vii, 3
Cosmological
  constant, 47
  horizon, 109
  parameters, x, 57
  principle, 3, 5
Current visibility radius, xiv, 108
Curvature
  constant, 8
  coordinates, 9
  tensors, 6

D
Dark energy, 56
Deceleration parameter, 66
Derivative of vector fields, vii
Doppler effect, 115

E
Einstein manifold, 4

S. Benenti, *Mathematical Foundations and Numerical Analysis of the Dynamics of an Isotropic Universe*, https://doi.org/10.1007/978-3-031-64033-9

Emission-reception relationship, 103
Emission time, 58, 102
Equation of state, 51
Events, 1
Expansion factor, vi

F
Factorization relation, 12
Fixed stars, 34
Foliation, 1
Free particle, 24

G
Galactic fluid, 1
Geodesic, 7
  equations, 7
Geometrical meaning, 6
Good ordering, vi
Gradient, 4

H
History, 1
Hubble
  factor, vi, 15
  horizon, 109
  law, vi, 1, 15
  length, 97
  parameter, 15
  radius, 97, 109
  time, 126
Hyper-spherical cap, 134

I
Inertial reference frame, 34
Inflection point, 66
Initial emission time, 106
Intermediate parameter, 119
Intrinsic meaning, 6
Inversion point, 63, 67
Invisibility radius of the Universe, 110
Isotropic
  tensor, 18
  vector, 18
Isotropy principle, 3

K
Key-events, 65
Key-times, xii
Killing-Hopf theorem, 10
Kinetic pressure, 49

L
Lagrangian coordinates, 17
Lambda cold dark matter model, v
Law of inertia, 34
Levi-Civita connection, 7
Linear connection, 6

M
Matter density, 55
Matter-radiation
  equilibrium, 55
  model, 56
Metric tensor, 7
MR-model, 56
Multiple zero, 62

N
Newtonian gravitazional constant, 47
Normalizaton time, 12

P
Parallel transport, 5
Particle horizon, 109
Peculiar velocity, vii, 36
Perfect fluid, 49
Photon, vii, 36
Photon decouplig, 101
Pointwise profile, xi
Primary data, x, xi
Primordial photons, 106
Principle of
  dimensional homogeneity, 15
  good ordering, v
  isotropy, 3
  simplicity, v
Profile
  of the Universe, xi, 64
Progression of a primordial photon, 107
Proper
  time, 45
  velocity, 45
Pseudo-cyclic profile, 66

R
Radial representation, 80
Radius of current visibility, 110
Reception time, 102

Recession speed, 15
Recombination, 101
Redshift, 117
  parameter, 118
Reference
  metric, 12
  space, 12
  time, vii, 12
Reference distance, 14
Reionization
  epoch, 101
  time, 108
Relative scalar velocity, 36
Relativistic cosmic connection, 37
Ricci
  curvature, 8
  scalar, 8
  tensor, 6
Riemann
  coordinates, 9
  form, 9
  tensor, 6
Riemannian manifold, 7

S
Scalar product, 7
Scale factor, vi, 12
Separation angle, 81
Simplicity, v
Simultaneous events, 3
Singular geodesics, 26
Slipher law, 125
Space vectors, 3
Spatial
  homogeneity, 4
  section, 1
  uniformity, 4
Special wandering particles, 21
Spectral shift, 117
Standard clock, 33
Standard model, 59
State parameter, 51
Stopping point, 62
Sub-luminal particle, 45
Super-luminal condition, 95
Super-luminal recession speed, 95
Symmetric connection, 6
Synchronous distance, 14

T
Total energy density, 51
Total internal pressure, 51
Totally covariant Riemann tensor, 8
Trans-luminal border, 96
Transport equations, 6
Transversal, 2

V
Visibility
  radius, 108
  relation, 108
Visible, 106

W
Wandering
  particles, vii, 20
Wandering particles
  special, 21
Weierstrass
  equation, xi, 62
  function, xi, 62
Weyl principle, vii, 5
World-line, 1

GPSR Compliance

The European Union's (EU) General Product Safety Regulation (GPSR) is a set of rules that requires consumer products to be safe and our obligations to ensure this.

If you have any concerns about our products, you can contact us on ProductSafety@springernature.com

In case Publisher is established outside the EU, the EU authorized representative is:

Springer Nature Customer Service Center GmbH
Europaplatz 3
69115 Heidelberg, Germany

Batch number: 10732090

Printed by Printforce, the Netherlands